A Sourcebook for

Environmental Education

A Practical Review based

on the Belgrade Charter

A Sourcebook for Environmental Education

A Practical Review based on the Belgrade Charter

Edited by

W. Leal Filho,
Z. Murphy
and K. O'Loan

UNIVERSITY OF
BRADFORD

MAKING KNOWLEDGE WORK

CRC Press
Taylor & Francis Group
Boca Raton London New York

CRC Press is an imprint of the
Taylor & Francis Group, an **informa** business

First Published 1996 Ertcee

Co-published with ERTCEE
Department of Environmental Science
University of Bradford

Published 2019 by CRC Press
Taylor & Francis Group
6000 Broken Sound Parkway NW, Suite 300
Boca Raton, FL 33487-2742

ISBN-13: 978-1-85070-768-4 (hbk)

Visit the Taylor & Francis Web site at
http://www.taylorandfrancis.com

and the CRC Press Web site at
http://www.crcpress.com

British Library Cataloguing in Publication Data

A sourcebook for environmental education: a practical review based on the Belgrade Charter
1. Environmental education
I. Filho, Walter Leal II. Murphy, Zena III. O'Loan, Kathryn
370.1'15

ISBN 1-85070-768-5

Typeset by European Research and Training Centre on Environmental Education

CONTENTS

Chapter 3

**CARING FOR THE EARTH: CASE STUDIES IN
ENVIRONMENTAL EDUCATION AND COMMUNICATION**   40
Frits Hesselink

Chapter 4

**ENVIRONMENTAL EDUCATION AT UNIVERSITY
LEVEL IN A DEVELOPING COUNTRY CONTEXT:
CURRENT TRENDS AND FUTURE CHALLENGES
IN GUYANA**   52
Paulette Bynoe

INTRODUCTION

The year of 1995 is a special year in the history of environmental education. For 20 years, since the 'Belgrade Workshop', one of the benchmark events in the history of environmental education which was jointly organised by UNESCO and UNEP, the International Environmental Education Programme (IEEP), launched in 1975, has been functioning.

The twentieth anniversary provided an ideal backdrop to examine the momentum begun at Belgrade. For this reason the European Research and Training Centre on Environmental Education (ERTCEE) of the University of Bradford, UK, hosted the 'Reviewing Belgrade Workshop'. Organised in liaison with UNESCO / UNEP and attended by active representatives from approximately 30 countries, the purpose of this event was to debate progress since Belgrade and, on the influential report produced there*.

This book, designed as a contribution to the twentieth anniversary celebrations of the IEEP, aims to provide an overview of some of the developments seen in the field of international environmental education and of the current contribution of leading institutions such as UNEP, UNESCO and IUCN. It also provides a new perspective into some of the most important areas for future action:

* teacher training
* environmental education at university level in a developing country context
* environmental education and communities in an African perspective
* environmental education projects
* networking and environmental education
* adult environmental education

In this book, readers will find a comprehensive review of the history of environmental education presented by Paul Pace (University of Malta), followed by a description of the achievements of IEEP by Wimala Ponniah (UNEP). A set of case studies in environmental education and communication is presented by Frits Hesselink (IUCN), while some trends and challenges on environmental education at university level in Guyana are described by Paulette Bynoe (University of Guyana). Joyce Glasgow (University of the West Indies) provides an overview of the provisions for environmental education in the training of teachers in a selection of countries, while Darlene Clover (International Council on Adult Education) describes the challenges in developing international environmental adult education. Christian Da Silva (IDRC) shares his experience with local environmental knowledge and environmental education in Tanzania, followed by Zena Murphy (ERTCEE) who discusses the role of networks in environmental education and Chris Taylor (ERTCEE) who debates the contribution of a project approach to the development of environmental education initiatives.

Following these chapters, David Uzzell (University of Sussex) and Richard Ballantyne (Queensland University of Technology) provide a checklist for the critical evaluation of informal experiences in environmental learning. Finally, some thoughts about the elements that need to be considered in furthering environmental education, along with some challenges, are presented.

The contributions provided by the various authors are based on their contacts over the years with individuals and institutions involved with environmental education at the formal and non-formal level and on their field experience in both industrialised and developing countries. It is thus hoped that the information amassed in this book will provide readers with an accurate view of the implications of the Belgrade Workshop in particular and of the process of evolution of environmental education as a whole, also providing some indications of the potential of some of the areas which are certain to be further developed in the late 1990s and beyond.

* *Reviewing Belgrade*. Report of the Intergovernmental Workshop Reviewing Belgrade, held at the University of Bradford, UK 23 to 28 July 1995. ERTCEE, University of Bradford, Bradford BD7 1DP, UK. ISBN 1 15843 140 3

Chapter 1

FROM BELGRADE TO BRADFORD - 20 YEARS OF ENVIRONMENTAL EDUCATION

Paul Pace

Department of Maths, Science & Technical Education
Faculty of Education
University of Malta
Msida MSD 06
Malta

INTRODUCTION

Humans have always been concerned about the environment and their role in it. However, this concern varies greatly amongst communities because of the wide spectrum of cultures, attitudes, values and beliefs. Thus, different human communities have treated the Earth in various ways: ranging from awe and respect to ruthless exploitation. In my opinion, our perception of Earth as our home, was the most important determining factor that influenced our concern for the environment. As the concept of a global village evolved, awareness of the dreary state of the environment and the far reaching effects of our activities emerged. This prepared the groundwork for a global effort to remedy the situation in order to ensure a future for all humanity.

PAVING THE WAY FOR BELGRADE

A major step forward in this emerging philosophy was the United Nations Conference on the Human Environment held in Stockholm, Sweden, 5-16 June 1972. Delegates from 113 Member States and observers from over 400 intergovernmental and non-governmental organisations attended this first world meeting on the state of the environment. Prior to this conference, the predominant conviction was that human activities and development were incompatible with the welfare of the environment. However, through its Declaration on Human Environment, the Stockholm Conference was soon to initiate a mechanism to change this.

The Declaration acknowledged the niche humans have in their environment as well as their responsibility for its care so as to ensure its availability to future generations. It also established that environmental problems transcend national frontiers (Robottom, 1987) and that further environmental damage could only be arrested with the mobilisation of international resources. This international awareness of the Earth as our common home "... placed environmental issues on the world's agenda" (Sandbrook, 1992) and prompted governmental and non-governmental organisations to widen their focus of activity (UNESCO-UNEP, 1988a).

However, the main feature that distinguished the event was the realisation that considerations of environmental improvement cannot be isolated from issues of development. The Conference was characterised by a conflict of interest between the wealthy industrialised countries (concerned with the end products of their indiscriminate development) and the poor Third World countries (desperate for any form of technology, even if polluting, to alleviate their major problem - poverty). Discussions clearly pointed out that underdevelopment was at the root of environmental degradation and that any attempt to improve the quality of the environment had to be coupled with an equal effort to better the quality of life. This very important conclusion, which was to characterise the main thrust of future environmental activities, could only be arrived at after a long series of activities (summarised in Figure 1) which preceded the Stockholm Conference and helped to shape opinions and to highlight world-wide priorities and needs.

The importance of a co-operative and international effort to promote environmental education was highlighted in 'Recommendation 96' of the Stockholm Conference. The Conference led to the establishment of UNEP (United Nations Environment Programme) which together with UNESCO (United Nations Educational, Scientific and Cultural Organisation) founded the UNESCO-UNEP International Environmental Education Programme (IEEP), launched in January 1975.

Although IEEP's actions have mostly focused on incorporating environmental education into the formal education systems of UNESCO Member States (Vinke, 1992) its original goals were much wider. IEEP was designed to:

a) facilitate the co-ordination and planning of international environmental education programmes;
b) promote international networking to facilitate the sharing of ideas and information;

Date	Event	Locality	Organised by	Remarks
29 Nov. - 4 Dec. 1965	Conference on Conservation of Nature and Natural Resources in South-East Asia	Bangkok, Thailand	IUCN in co-operation with WWF	focused on promoting environmental education in schools, public education and training in conservation (UNEP, 1976a)
4 - 13 Sept. 1968	Biosphere Conference	Paris, France	UNESCO with assistance from UN, FAO, WHO, IUCN and the International Biological programme	stressed the need for a global awareness of environmental problems and the production of educational resources on the environment (Sterling and Cooper, 1992)
20 June - 11 July 1970	International Working Meeting on Environmental Education in the School Curriculum	Carson City, Nevada, USA	IUCN in co-operation with UNESCO	principally noted for coining an official definition for the term 'environmental education', the latter being widely accepted (Sterling and Cooper, 1992)
9 - 12 Feb. 1970	First European Conservation Conference	Strasbourg, France	Council of Europe	it was the first time that EE was introduced in European conservation strategies (Docter Institute for Environmental Studies, 1991)
5 - 8 April 1971	Workshop on Environmental Education at the University Level	Tours, France	CERI (Centre for Education Research and Innovation), OECD	identified the ideological base of EE at university level and the need and urgency for a new EE programme at university level (CERI, 1973)
15-18 Dec. 1971	European Working Conference on Environmental Education	Rushlilikon, Switzerland	IUCN in co-operation with WWF	Specific recommendations for EE projects and programmes for primary and secondary levels, teacher training, out-of-school and higher education were made (UNEP, 1976a)

Figure 1. Activities which prepared the foundations for the Stockholm Conference

c) co-ordinate research in teaching and learning;
d) formulate and evaluate new approaches, materials and programmes for environmental education at the formal and informal levels of education and for all ages;
e) train and retrain personnel to manage environmental education programmes, and
f) provide advisory services relating to environmental education.

Research, experimentation and development of innovations in the field of environmental education are the three principal areas which characterised IEEP activities. This was primarily achieved through a world-wide programme of pilot projects (see Figure 2). Furthermore, IEEP sought to gather, organise and disseminate information on environmental education principally through a global network system of information about resource persons and institutions related to environmental education and the publication of *'Connect'* - an international newsletter promoting discussion and the elaboration of policies and strategies for the development of environmental education at local, national, regional and global levels (UNEP, 1977). IEEP's initial 3-year schedule was divided into 3 action phases:

Phase 1: The preparation of documents for the international Belgrade Workshop. The documentation amassed included (a) the results of an internationally administered, questionnaire-based needs survey (revealing the need for a great global effort to improve environmental education at all levels of education), (b) reports from the 'consultants' missions' to developing UNESCO member states which assessed the state of environmental education in the country and investigated the possibilities for future action, and (c) commissioned 'trend-papers' describing current environmental education practices from various world regions.

Phase 2: The convening of the international Belgrade Workshop to discuss the documents and make recommendations for a global environmental education action programme. These recommendations served as a working document for a series of regional seminars which would review, evaluate and modify the recommendations on the basis of regional needs and characteristics. These regional recommendations would in turn become the working document for the Intergovernmental Conference in Tbilisi whose focus would be the drawing up of recommendations for governments on policies for the implementation of environmental education programmes.

Region	Country	Project
Africa	Ghana	Development of multimedia materials on environmental science education components, for use by primary school teachers in Sub-Sahara Africa
	Kenya	The Kiambu High School project for the development of an integrated environmental science education programme for secondary schools
	Senegal	Environmental education pilot project for the adult populations of Africa
Arab States	Jordan	Pilot project on water pollution and purification in the Ghore area of the Jordan Valley
	Egypt	Environmental education in youth clubs and associations
Asia and Oceania	Afghanistan	Design, development and implementation of instructional materials in environmental education for primary schools
	Indonesia	An environmental education experiment using an integrative approach in Jakarta primary schools
	Mongolia	Non formal environmental education for the general public and for specific social groups
Latin America and the Caribbean	Colombia	Educational programme for the conservation of the ecological balance and renewable natural resources of Colombia
	Guatemala	Pilot project to organise the first four years of primary education around environmental problems
	Panama	Environmental education as a component of the work study programmes of general basic education centres
	Peru	Experimental environmental education project for adaptation in the third year programme of Peruvian secondary education
	Venezuela	Environmental education for secondary schools
Europe and North America	Ukrainian SSR	Methodological interdisciplinary research concerning the integration of school and society in the sphere of environmental education
	Canada-USA	A two-nation regional environmental education programme based on the shared marine environment of the Pacific Northwest
	USA-Canada	An international environmental resources network (Internet)
	France	Interdisciplinary methodological research concerning the acquisition of knowledge, values, attitudes and skills related to the environment by secondary school students
	United Kingdom	In-service training of European educators in urban environmental studies and research

Phase 3: The IEEP was to be involved principally in the implementation of the recommendations of the Intergovernmental Conference. However efforts still continue to strengthen further networking between nations, continue research on the development and evaluation of environmental education initiatives, provide on-going support to promote environmental education programmes and disseminate publications on environmental education.

BELGRADE

From 13-22 October 1975, as part of the IEEP, UNESCO-UNEP convened the International Environmental Education Workshop in Belgrade, Yugoslavia. The workshop was attended by around 100 educational specialists from 64 countries who gathered together with the aim of evaluating trends in environmental education and to arrive at a set of recommendations addressing appropriate target groups for an international effort to promote environmental education. These recommendations formed the 'Belgrade Charter - a global framework for environmental education' which was the first intergovernmental statement (Scottish Office Environment Department, 1993) which outlined the broad philosophy and principles of environmental education (Sterling and Cooper, 1992). The very first issue of *Connect* defined the Belgrade Charter as a historic document which " ... laid down the principles and established the guidelines for the world-wide environmental education of a generation which spans the earth" (UNEP, 1976a).

Reiterating Stockholm's concerns about the rapid development of humankind and its negative effects on the environment, the Belgrade Charter acknowledges the need for a collective effort to improve (on a global level) the quality of life and the quality of the environment together. The Charter rejects the fragmentary approach to environmental problem solving and advocates a 'new global ethic' necessitating a change in priorities and behaviour of individual citizens. This global change in behaviour and value systems can only be achieved successfully through environmental education addressing, through formal and informal channels, all the different sectors of the population. Although notions of environmental education had been maturing for quite some time (see Figures 1, 3a and 3b) the Belgrade Workshop was the first meeting of its type to clearly identify its aims, objectives and guiding principles.

Date	Event	Locality	Organised by	Remarks
22-28 June 1972	International Seminar on Environmental Conservation Education in the School Curriculum in East-European Countries	Kroscienko, Poland	IUCN	aimed to further develop the Rushlikon Conference of Dec. 1971 (UNEP, 1976 a)
5-17 Sept. 1972	International Workshop on Environmental Studies in Higher Education and Teacher Training	London, Canada	The North American Committee in co-operation with IUCN	highlighted problems facing EE in higher education, especially concerning teacher and professional training (UNEP, 1976a)
4 - 16 March 1974	Eastern Africa Youth Meeting on Environmental Conservation	Nairobi, Kenya	UNESCO, WWF, International Youth Federation & African Wildlife Leadership Federation	aimed at promoting an understanding of and active involvement in environmental conservation (UNEP, 1976a)
28 April - 2 May 1974	International Working Meeting on Environment in Educational Programmes	Cairo, Egypt	the Arab League Cultural and Scientific Organisation (ALESCO) and UNEP	considered an EE plan for primary and secondary levels, and a programme of production of technical manuals, guides and source books addressing the specific problems and needs of the Arab States (Leal Filho1990)
4-7 June 1974	Conference on Environmental Education at Post-secondary level: Review of Experience - Future Action	Rungsted, Denmark	CERI (Centre for Education Research and Innovation), OECD	analysed case studies from OECD countries, made recommendations on EE courses tertiary level for generalists and specialists, educators, decision-makers and members of professions concerned with the environment (CERI, 1974)

Figure 3a. Events following the Stockholm Conference and preceding the Belgrade Workshop

Date	Event	Locality	Organised by	Remarks
27 Aug. - 4 Sept. 1974	Pilot Seminar on Environmental Education Methodology in East Africa	Mombasa, Kenya.	UNEP, IUCN and the World Confederation of Organisations of the Teaching Profession (WCOTP)	aimed to bring together experts in EE, education and educational technology to discuss the inclusion of environmental issues in the teaching systems of East African countries (Leal Filho, 1990)
16-25 Oct. 1974	Seminar on Education for the Conservation of Renewable Natural Resources in High School Curricula	Cordoba, Argentina	Centre of Ecology and Renewable Natural Resources, National University of Cordoba in co-operation with U.S. Forest Institute for Ocean and Mountain Studies, Argentine Secretary of State for Natural Resources and the Human Environment, WWF and IUCN	established guidelines for the introduction of EE programmes in high schools and made a series of recommendations on curriculum content, pre-service teacher training and extra-curricular activities (UNEP, 1976a)

Figure 3b. Events following the Stockholm Conference and preceding the Belgrade Workshop

During the Workshop itself, plans for follow-up activities were already underway. Regional groups met and prepared for the regional meetings of environmental education experts. The goal set for these meetings was that of examining the Belgrade Charter recommendations and to adapt them to regional needs. During these regional meetings (see Figure 4), educational experts (a) identified and discussed regional environmental education activities, (b) revised the Belgrade guidelines in the light of regional environmental education needs, (c) promoted the exchange of ideas and information in order to strengthen regional networks and co-operation in environmental education, and (d) formulated strategies for further environmental education action at the regional level. However another aim that was achieved during these meetings was the orientation of decision-makers, curriculum planners and experienced educators towards environmental education concepts and methodologies, thus preparing the way to the next great event - the Tbilisi Intergovernmental Conference (UNEP, 1976b).

Figure 4. List of the Regional Meetings on Environmental Education following the Belgrade Workshop (UNEP, 1977).

Date	Region	Meeting Place	No. of Participants *	No. of Countries Represented
11-16 Sept. 1976	Africa	Brazzaville, People's Republic of the Congo	27	12
15-20 Nov. 1976	Asia and Oceania	Bangkok, Thailand	24	17
21-25 Nov. 1976	Arab States	Kuwait	30	13
24-30 Nov. 1976	Latin America and the Caribbean	Bogota, Colombia	30	19
27-31 Jan 1977	Europe and North America	Helsinki, Finland	30	23

* not counting observers

9

FOLLOWING BELGRADE

UNESCO, in co-operation with UNEP, convened the Intergovernmental Conference on Environmental Education at Tbilisi, Georgia, 14 - 26 October 1977. It was attended by delegates from 66 UNESCO member states and 20 international NGOs as well as various observers from non-member states and UN agencies and programmes. In all 265 delegates and 65 representatives and observers took part in the Tbilisi Conference (UNEP, 1978a). While the Belgrade Workshop was characterised by the participation of education experts, Tbilisi's target group was policy-makers. This helped to provide a more formal intergovernmental recognition of the environmental education concept developed earlier in Belgrade (Robottom, 1987). The Tbilisi Conference set out to define environmental education, to formulate its goals and objectives and to evolve strategies for its development.

At Tbilisi the delegates accepted and elaborated the goals, objectives and principles of environmental education outlined in the Belgrade Charter. They recommended that environmental education should be directed to the general public of any age as well as specific occupational and social groups through both formal and informal education. Tbilisi also encouraged the setting up of environmental education programmes in various countries based on their particular needs and situations. In fact IEEP initiatives intensified following the Tbilisi Conference. Action was initiated to promote environmental education by:

a) its systematic integration into national education policies;
b) the promotion of interdisciplinary and inter-institutional co-operation;
c) the promotion of interdisciplinarity and a problem-solving methodology in curricula and educational material;
d) the production of multimedia programmes for all;
e) the planning of research and experimentation on educational processes and its dissemination;
f) the incorporation of the environmental education dimension in pre- and in-service training of educators;
g) the provision of financial and technical support for a number of national training activities of key personnel;
h) the strengthening of international, national and regional co-operation among governmental and non-governmental organisations, and
i) the assignment of environmental education consultants for each of the four world regions (UNESCO-UNEP, 1988a).

Figure 5 summarises the environmental education initiatives that have been officially recognised as being follow-up activities of the Tbilisi Conference.

Another immediate follow up of the Tbilisi Conference was an international workshop on co-operation among NGOs in the field of environmental education which took place in Geneva in October 1977. This was the result of a joint initiative between UNEP and the Environment Liaison Centre (ELC), which is a focus for international NGO liaison. Participants who had attended the Tbilisi Conference now focused their attention on making recommendations for ELC's post-Tbilisi action. The meeting proposed the setting up of a central body to facilitate NGO work in environmental education at a regional level. This central body would be responsible for the dissemination of information about the availability of environmental education resources; the co-ordination of their distribution and the recording of evaluations of the resources made by user NGOs (UNEP, 1978c).

In June 1978, a three-day meeting (21-23 June) was convened by UNEP in UNESCO's Paris headquarters. The meeting was attended by representatives from the Economic and Social Commission for Asia and the Pacific; the Food and Agriculture Organisation; the International Labour Organisation, the International Maritime Consultative Organisation, the World Health Organisation, the World Meteorological Organisation, the United Nations Fund for Population Activities, UNESCO and UNEP. The discussion focused on a working paper prepared by UNESCO in consultation with UNEP which addressed three major issues: (a) the development of environmental education in the light of the Tbilisi Conference; (b) the current environmental education and training activities of UN organisations; and (c) the possibility for planned co-operative action, by the participating organisations, to implement the Tbilisi recommendations. The agreed plans for future action included the integration of environmental education into co-operative UN activities; the exchange of information and experiences in environmental education; the development of environmental education programmes and resources; the incorporation of an environmental dimension in pre- and in-service training of education personnel; and the promotion of research, experimentation and evaluation in environmental education (UNEP, 1978c).

Towards the end of 1978, a principal area of activity of the IEEP was set into operation as a direct follow up to the Tbilisi Conference. Regional (and sub-regional) Workshops in Environmental Education for key personnel (see Figure 6) were organised targeting teacher educators, curriculum developers,

Country / Region	Initiative
United States	The 1978 National Leadership Conference on Environmental Education held in Washington, D.C., USA (28-30 March). Participants reviewed the recommendations of past conferences - particularly the Tbilisi Conference - and developed specific action recommendations for the implementation of a national strategy
Union of Soviet Socialist Republics	On 13 April 1978 the State Committee for Science and Technology under the USSR Council of Ministers approved a plan for developing education in the field of environmental protection and improvement in the light of the Tbilisi recommendations.
Federal Republic of Germany	Conference held in Munich (26-28 April 1978) which specifically discussed the results and recommendations of the Tbilisi Conference and worked out plans for their implementation.
United Kingdom	The national seminar "Response to the Challenge of Tbilisi" was convened on 27 June 1978 with the aim of reviewing the Tbilisi recommendations, to discuss and identify specific problem areas of EE in the UK and what actions can be undertaken to surpass them.
Poland	A seminar with the theme: 'University, Environment and Society' was organised at Warsaw (26-29 June 1978). Discussions focused on the recommendations of the Helsinki Treaty (1975) as well as those of the Tbilisi Conference relative to the role of universities in the face of the environmental challenge.
Czechoslovak Socialist Republic	A national conference was organised in the Krkonose National Park of Bohemia, Svaty Petr (26-29 September 1978). It dealt with the implementation of the Tbilisi recommendations on EE within both formal and non formal educational systems.
West Africa	Seminar on the development of instructional materials for the environmental education of adults in West Africa was held in Dakar, Senegal (25 September- 6 October 1978). The seminar was the first phase of a pilot project whose second phase involved the testing of the instructional materials in Senegalese villages.
Kenya	Workshop on Environmental Education and Training in African Universities held in Nairobi (3-9 December 1978). The objectives were to further EE training and research at universities through action-oriented programmes and by establishing networks and to train personnel in the principles and practice of EE.
Poland	Preparation and publication of a manual on the natural environment for teachers and students of all disciplines (1978). In February 1979 the current state of EE in Poland was discussed at a meeting on EE. Recommendations for policies for the future were drawn.
Bulgaria	In June 1979, the Ministry of Education organised a national training workshop in EE which was attended by teachers, administrators of teacher-training institutes, researchers and officials responsible for nature conservation.
United Kingdom	A second national seminar with the theme 'Response to the Challenge of Tbilisi' was convened on 6 July 1979 with the aim of reviewing any progress made from the first seminar (June 1978), identifying further needs, planning further courses of action and deciding on the UK's contribution to international EE

Figure 5. Post-Tbilisi Initiatives at the National Level (UNEP, 1978c; UNEP, 1978d; UNEP, 1979b; UNEP, 1979c;

Figure 6. Regional (and Sub-Regional) Workshops in Environmental Education for Key Personnel (UNEP, 1979a; UNEP 1980a; UNEP, 1980b; UNEP, 1980c; UNEP, 1981a; UNEP, 1981b; UNEP, 1981c)

Region	Date	Locality	Number of Countries Represented
Africa	11-20 December 1978	Dakar, Senegal	20
Latin America	29 Oct. - 7 Nov. 1979	San Jose, Costa Rica	20
The Caribbean (Sub-regional workshop)	9-20 June 1980	Antigua	10
Asia and Oceania	22-29 September 1980	Bangkok Thailand	17
Europe (Sub-regional workshop)	13-17 October 1980	Prague and Hluboka, Czechoslovak Socialist Republic	9
Europe - North America	8-12 December 1980	Essen, Federal Republic of Germany	20
Arab States	12-19 May 1981	Manama, Bahrain	10

supervisors, educational planners and administrators. The workshops focused on the development of strategies establishing:

a) environmental education programmes at the national level;
b) curriculum development and the preparation of teaching/learning materials for the primary and secondary schools and teacher-training institutions;
c) training of teachers in environmental education for primary and secondary schools and teacher education institutions (UNEP, 1979a).

Following this 'post-Tbilisi mania' of activities generated by UN organisations, matters concerning the implementation of the Tbilisi recommendations seemed to calm down. This may probably be because attention was shifted to the development of 'smaller' environmental education activities conducted at local and national levels. However, from the major international activities that followed Tbilisi, it is apparent that the world was undergoing an acclimatisation period. Countries were probably not ready to adopt the radical changes in life style implied by the recommendations. Whatever the reason, it seems that the challenges set by Tbilisi have not really been met (Sterling, 1992).

Another major step in the development of a new environmental ethic was the publication of the World Conservation Strategy (WCS) by WWF, UNEP and IUCN in co-operation with UNESCO and FAO in 1980. Launched simultaneously in 30 countries, the document was the result of 2 years work by 450 government agencies and more than 700 experts from 100 different countries. It was chiefly aimed at government policy makers, conservationists, and developers. The aim of WCS was "... to help advance the achievement of sustainable development through the conservation of living resources" (IUCN, 1980). To achieve this aim WCS sets 3 objectives: (a) the preservation of ecological processes and life-support systems; (b) maintain genetic diversity; and (c) ensure the sustainable use of species and ecosystems (IUCN, 1980). After reviewing current environmental problems, WCS outlines its objectives and proposes effective national and international actions to achieve them. The strategy highlights the importance of conserving the living resource to ensure human survival. In this context, it proposes the concept of sustainable development and identifies the priority conservation issues and the main requisites for dealing with them. WCS also considers aspects of training, research, participation, and education.

In April 1987 the United Nations World Commission on Environment and Development (WCED), made up of 22 persons from 22 countries, produced a 380 page report called 'Our Common Future' (better known as the 'Brundtland Report'). Based on nearly 3 years of research involving public hearings all around the world and data gathering from all sectors and strata of society, the report was intended as a progress report on the Stockholm Conference. Its main contribution was that, after reviewing the critical environmental conditions of our planet, it concludes that economic growth cannot be arrested, but it can be approached differently. The report urges governments to make environmental concerns central issues in their decision-making processes to ensure that meeting the needs of the present citizens does not compromise the ability of future generations to meet their own needs, i.e. sustainable development.

The UN General Assembly in resolution 42/187 welcomed and decided to transmit the report "Believing that sustainable development ... should become a central guiding principle of the UN, governments and private institutions, organisations and enterprises ..." (UNEP, 1988). Moreover it emphasised that sustainable development is "... an essential prerequisite for the eradication of poverty and for enhancing the resource base on which present and future generations depend ..." (UNEP, 1988). This reflects the two aspects of sustainable development. For rich countries sustainable development is essentially translated in issues of whether to adopt new 'environmentally-friendly' policies like recycling, alternative energy sources, and conservation of landscapes. On the other hand, sustainable development for poor countries involves issues of social justice, the generation of wealth and its fair redistribution (Sandbrook, 1992). In this framework environmental education is once again acknowledged as a critical tool to achieve sustainable development.

'Our Common Future' had an important development. The WCED chairperson, Ms Gro Harlem Bruntland, became Norway's Prime Minister shortly after working on the report. This enabled her to adopt the report's suggestions at the highest level of decision-making in her country. This initiative started what Sandbrook (1992) calls an 'environmental beauty contest' among heads of state who sought to put the Bruntland Report's suggestions into practice - proving once again the strong hold environmental issues have gained on political agendas. Developments led 50 nation leaders to demand a major event to discuss and act on the report and in the resulting debate at the UN in 1989, resolution 44/228 was passed stating that there should be a UN Conference on Environment and Development and determined the ground it would cover (Sandbrook, 1992).

The issues of sustainable development, responsible use of resources and habitat rehabilitation were also treated in another report commissioned and approved by the Governing Council of UNEP in May 1987. The report, entitled: 'Environmental Perspective to the Year 2000 and Beyond', was adopted by the UN General Assembly in resolution 42/186. The report provided governments with a broad framework guiding national action and international co-operation on environmentally sound development policies and programmes. It was also seen as a guide to the preparation of further UN system-wide medium term environment programmes and the medium term programmes of UN organisations and bodies (UNEP, 1988).

Ten years after the Tbilisi Conference, UNESCO-UNEP organised the International Congress on Environmental Education and Training, in Moscow (17-21 August 1987), which was also appropriately termed 'Tbilisi plus ten'. After a period of relative inactivity in the field of environmental education, the need for more action and less talk was becoming more urgent. While endorsing the Tbilisi recommendations and the action suggested by 'Our Common Future', the 300 educators from 80 countries who attended the Congress reiterated the need for environmental education at all levels of education. Once again the major theme was sustainable development (UNEP, 1987).

During the congress, three commissions considered the major aspects of environmental education: (a) environmental education and training of teachers for school and out-of-school activities, (b) environmental education and training in general university education, and (c) specialised environmental training. The congress also organised five symposia focusing on the following themes:

1) international understanding and environmental problems: the role of environmental education;
2) environmental education and training: their contribution in the perspective of sustainable socio-economic development;
3) role of media and new communication systems in the promotion of environmental education and information;
4) role of biosphere reserves and other protected areas in the dissemination of ecological knowledge and training of ecological specialists, and
5) national experiences and the contribution of NGOs in the development of environmental education and training (UNEP, 1987).

It was obvious, from the contributions made, that the introduction of environmental education in schools had not been the bed of roses it had seemed, mainly because it required a radical reconsideration of the rationale of the curriculum and a reorganisation of teaching methods (Booth, 1987).

The end product of these discussions was the drawing up of an 'International Strategy for Action in the Field of Environmental Education and Training for the 1990s' which, in an attempt to reverse the state of environmental illiteracy, dedicated the 1990s as the 'World Decade for Environmental Education' (Leal Filho, 1990). The aims of the international strategy are concerned with: (i) the search for and implementation of effective models of environmental education, training and information; (ii) a general awareness of the causes and effects of environmental problems; (iii) a general acceptance of the need for an integrated approach to solving these problems;

and (iv) the training, at various levels, of the personnel needed for the rational management of the environment in view of achieving sustainable development at community, national, regional and world-wide levels. The main headings for discussion were: 'information access', 'research and experimentation on content and methods', 'educational programmes and resources', 'training of personnel', 'technical and vocational education', 'educating and informing the general public', 'general university education', 'specialist training', and 'international and regional co-operation' (UNESCO-UNEP, 1988b).

Once again, following the Moscow Congress, progress in environmental education was taken up at the local and regional levels. Various nations and world regions studied and adopted plans to include environmental education in their strategies to achieve a better relationship with the environment. Such 'small scale' activities are important because they allow the assimilation of proposed recommendations made at international gatherings and slowly help different people to acclimatise to the changes proposed. However a regular input of international meetings and initiatives is also required to prevent local initiatives from grinding to a halt and to underline the important fact that although activities at the local level may vary because of different cultures and traditions, the end goal should be a common one. This feeling of belonging to the same home is a necessary requisite to generate the attitudes and values required for the environmental ethic which will ensure the survival of humanity. In order "... to secure a widespread and deeply-held commitment to a new ethic, the ethic for sustainable living, and to translate its principles into practice ..." and "... to integrate conservation and development: conservation to keep our actions within the Earth's capacity, and development to enable people everywhere to enjoy long, healthy and fulfilling lives ..." IUCN published 'Caring for the Earth: a Strategy for Sustainable Living' (IUCN/UNEP/WWF, 1991).

Extending and emphasising the principles set by the World Conservation Strategy, 'Caring for the Earth' identifies and is structured around nine principles for sustainable living:

1) respect and care for the community of life;
2) improve the quality of human life;
3) conserve the Earth's vitality and diversity;
4) minimise the depletion of non-renewable resources;
5) keep within the Earth's carrying capacity;
6) change personal attitudes and practices;
7) enable communities to care for their own environments;

8) provide a national framework for integrating development and conservation; and

9) create a global alliance (IUCN/UNEP/ WWF, 1991).

Another important development highlighted in 'Caring for the Earth' is the audience for which it is written. Previous plans directed at environmental action - including the 'World Conservation Strategy' itself - were traditionally addressed to governments and other policy makers. However this document, when considering " ... those who shape policy and make decisions that effect the course of development and the condition of our environment ...", intends to include every individual and not just people in 'high places' (IUCN/UNEP/WWF, 1991). Rather than emphasising the traditional top-down approach, 'Caring for the Earth' addresses the general public and invites them to take concrete actions towards sustainable living. This approach, which renders the population active participants in environmental action rather than passive recipients of directives, characterised the development of the next great international development - the United Nations Conference on Environment and Development (UNCED).

The Rio 'Earth Summit' (3 - 14 June 1992) called upon governments to direct strategies for integrating environment and development into education at all levels by the next three years so as to set up a world-wide programme to develop environmental and developmental literacy by the year 2000 as the learning requirement for an environmentally competent earth citizenry. "...Agenda 21 constitutes the most comprehensive and far-reaching programme of action ever approved by the world community. And the fact that their approval was at the highest political level lends it special authority and importance. For the first time in international politics we have consensus that the future of the planet is at stake if we do not reverse the process of abusing it" (Strong, 1992).

Although principally addressing governments, UNCED was an appeal to end the fragmentary method of organising environmental action. It recognised the responsibility and the role every individual citizen plays in the struggle to find a balance between quality of life and quality of the environment. It was a call to get all the interested parties working together on the planning and implementation of strategies for sustainable living.

This popularisation of the decision-making processes in matters concerning the environment and development is clearly spelled out in Principle 10 of the 'Rio Declaration on Environment and Development'. It declares that "Environmental issues are best handled with the participation of all concerned citizens, at the relevant level. At the national level, each individual shall have appropriate access to information concerning the environment that is held by public authorities, including information on hazardous materials and activities in their communities, and the opportunity to participate in decision-making processes. States shall facilitate and encourage public awareness and participation by making information widely available. Effective access to judicial and administrative proceedings, including redress and remedy, shall be provided" (UNCED, 1992). Since education is "linked to virtually all areas in Agenda 21" it is quite obvious that UNCED approved Chapter 36 on 'Promoting Education, Public Awareness and Training' as part of Agenda 21.

However, it is a well known fact that the UNCED process was largely characterised by discussion on the major environmental issues. With the threat of the Cold War now gone from the world political agenda, the attention at Rio now focused on the rift between the rich and the poor countries. While the rich countries were concerned with environmental issues, poor countries insisted that development should form an important part of the discussion on environment. Sandbrook (1992) feels that the major focus in Rio was not the environment, but the economy and how the environment affects it. This concern is clearly evident particularly in Principle 1 ("Human beings are at the centre of concerns for sustainable development. They are entitled to a healthy and productive life in harmony with nature") and Principle 3 ("The right to development must be fulfilled so as to equitably meet developmental and environmental needs of present and future generations") of the Rio Declaration (UNCED, 1992).

Hence, because of these politically 'hot' issues, the question of global environmental education was largely overlooked. With the aim of resuming and furthering the debate on international environmental education Dr Walter Leal Filho and Monica Hale convened an International Workshop on Environmental Education with the theme 'Promoting International Environmental Education', at Rio de Janeiro, Brazil, 4-6 June 1992, parallel to the Earth Summit. Representatives from 33 countries attended the meeting to communicate the developments and status of environmental education in their countries, to exchange views, identify problems of implementation and to motivate further actions in this field on an international level. As a follow up to the meeting, regional workshops were planned between 1992-95 during which delegates would have the opportunity of examining trends and problems associated with environmental education common to the region in question, seek solutions based on existing experience, and promote environmental education projects drawing from local and regional expertise (Leal Filho and Hale, 1992).

In the meantime, UNESCO and the International Chamber of Commerce in co-operation with UNEP, initiated the first UNCED follow-up in Toronto, Canada, September 1992, with a World Congress for Education and Communication on Environment and Development (ECO-ED). The purpose of ECO-ED was to stimulate informed action by improving the accuracy, quality and delivery of education and communication relating to the environment and sustainable development. This practical action-oriented initiative, promoted the exchange of relevant knowledge among educators, scientists, representatives of labour, business, governments, voluntary organisations and the media as well as educators, students, and indigenous people (UNEP, 1992).

ARRIVING AT BRADFORD

Presently we are characteristically re-experiencing another 'assimilation period' after the onslaught of the Rio Earth Summit. Various nations and regions are re-evaluating their provisions for environmental and development education. In my journey from Belgrade to Bradford I will conclude by citing just a handful of activities occurring in Bradford which, besides reflecting the current key issues in environmental education, serve to illustrate the type of environmental education activity going on at present in various world regions.

The University of Bradford, UK hosted the Commonwealth Conference on Environmental Education, 18-23 July 1993, attended by 80 delegates from around 40 Commonwealth countries as a contribution to the IEEP. The aim of the conference was to catalyse and promote the systematic development of environmental education programmes in Commonwealth countries. The event also provided a forum for the sharing of ideas and experiences, thus stimulating contacts with the aim of creating networks and working links between the participating countries. As a concrete follow-up of the conference, the delegates drew up the 'Bradford Declaration on Environmental Education in the Commonwealth'. The document asks heads of government of Commonwealth states to:

a) acknowledge the importance of environmental education in all efforts to achieve sustainable development;
b) develop local environmental education strategies,
c) to introduce and support environmental education programmes at all levels of formal and informal education, and
d) to create links to promote co-operation and maximise resources and funds.

The document was brought to the attention of the Commonwealth Ministers of Education meeting in 1994 so that it could be discussed in the 1995 Heads of Government meeting (Leal Filho, 1993).

Another meeting, organised by the European Research and Training Centre on Environmental Education (ERTCEE) of the University of Bradford, was the Colloquium on Environmental Education in Europe, 25-27 February 1994. Representatives from 11 European Countries (including delegates from eastern Europe) highlighted the common trends of European environmental education and its allied problems. During a plenary meeting participants agreed to set up a Study Group on Environmental Education in Europe (SGEEE) with the aim of congregating European environmental education specialists to oversee developments in the field of environmental education in Europe. Over 700 individuals and institutions have joined SGEEE which publishes a newsletter (also available on e-mail) with the aim of exchanging information about environmental education activities in different areas so as to promote co-operation and forge links. The colloquium also drew up a declaration: 'Towards an Environmentally Responsible Europe', requesting the Commission of the European Communities, Council of Europe, individual governments as well as national and international agencies to promote environmental education by providing suitable logistical and financial support for its incorporation in school curricula and in teacher training programmes (Leal Filho, 1994).

A final activity worth mentioning in my itinerary is the International Workshop on Environmental Journalism, organised by ERTCEE also at Bradford, 28-30 April 1995. Editors, reporters and freelance journalists from various countries could experience through case studies and hands-on experiences how electronic media, and especially networks could be used to promote environmental journalism. The discussion was principally focused on discovering efficient ways of equipping the public with the information required to enable them to make better informed decisions about the environment and development.

CONCLUSION

Environmental education has evolved quite rapidly from a mere interest of a handful of people to a crucial item in the political agenda of a large number of countries. Personally I think that there are two possible reasons accounting for its fast evolutionary rate. Firstly, environmental education addresses one of the major issues crucial to our survival on the planet, and secondly because the message being advanced is not one of impending disaster but one of hope. People are beginning to realise that the world can be improved through collective and co-operative social action. The dream that quality of life and quality of the environment are compatible can finally start to materialise.

REFERENCES

Booth, R. (1987) 'Thoughts after Moscow'. In CEE (Council for Environmental Education) *AREE - Annual Review of Environmental Education*. No.1 - Review of 1987. CEE, Reading, UK.

CERI (Centre for Education Research and Innovation) (1973). *Environmental Education at University Level: Trends and Data*. Organisation for Economic Co-operation and Development (OECD), Paris, France.

CERI (Centre for Education Research and Innovation) (1974). *Environmental Education at Post Secondary Level: Courses for Educators, Decision-Makers and Members of Professions concerned with the Environment Volume 2*. Organisation for Economic Co-operation and Development (OECD), Paris, France.

Docter Institute for Environmental Studies (1991) *European Environmental Yearbook*. Docter International London, UK.

IUCN (1980) *World Conservation Strategy: Living Resource Conservation for Sustainable Development*. IUCN, Gland, Switzerland.

IUCN/UNEP/WWF (1991) *Caring for the Earth. Strategy for Sustainable Living*. Gland, Switzerland.

Leal Filho, W.D.S. (1990) *Environmental Education for a Developing Country*. Unpublished Ph.D. Thesis, University of Bradford.

Leal Filho, W.D.S. (Ed) (1993) *Priorities for Environmental Education in the Commonwealth*. Proceedings of the International Conference 'Environmental Education in the Commonwealth'. University of Bradford, Bradford, UK.

Leal Filho, W.D.S. (Ed) (1994) *Trends in Environmental Education in Europe*. Proceedings of the Colloquium 'Trends in Environmental Education in Europe'. University of Bradford, Bradford, UK.

Leal Filho, W.D.S. and Hale, M. (1992) *Promoting International Environmental Education*. Proceedings of the 'International Workshop on Environmental Education'. Rio de Janeiro, Brazil, 4-6 June 1992.

Robottom, I. (1987) 'Towards inquiry-based professional development in environmental education'. In Robottom, I. (Ed) *Environmental Education: Practice And Possibility*. Deakin University Press, Victoria, Australia.

Sandbrook, R. (1992) 'From Stockholm to Rio'. In Quarrie, J. (Ed) *Earth Summit '92*. The Regency Press Corporation, London.

Scottish Office Environment Department (1993). *Learning for Life - A National Strategy for Environmental Education in Scotland*. HMSO, Edinburgh.

Sterling, S. (1992) 'Mapping environmental education - principles, progresses and potential'. In Leal Filho, W. and Palmer, J.A. (Eds.) *Key Issues in Environmental Education*. The Horton Print Group, UK.

Sterling, S. and Cooper, G. 1992. *In Touch: Environmental Education for Europe*. WWF, UK.

Strong, M. (1992) 'Forward'. In Quarrie, J. (Ed) *Earth Summit '92*. The Regency Press Corporation, London.

UNEP (1976a) *Connect* Vol. I, No. 1. January 1976.

UNEP (1976b) *Connect* Vol. I, No. 2. April 1976.

UNEP (1977) *Connect* Vol. II, No. 1. March 1977.

UNEP (1978a) *Connect* Vol. III, No. 1. January 1978.

UNEP (1978b) *Connect* Vol. III, No. 2. May 1978.

UNEP (1978c) *Connect* Vol. III, No. 3. September 1978.

UNEP (1978d) *Connect* Vol. III, No. 4. December 1978.

UNEP (1979a) *Connect* Vol. IV, No. 1. March 1979.

UNEP (1979b) *Connect* Vol. IV, No. 2. June 1979.

UNEP (1979c) *Connect* Vol. IV, No. 3. September 1979.

UNEP (1979d) *Connect* Vol. IV, No. 4. December 1979.

UNEP (1980a) *Connect* Vol. V, No. 1. March 1980.

UNEP (1980b) *Connect* Vol. V, No. 3. September 1980.

UNEP (1980c) *Connect* Vol. V, No. 4. December 1980.

UNEP (1981a) *Connect* Vol. VI, No. 1. March 1981.

UNEP (1981b) *Connect* Vol. VI, No. 2. June 1981.

UNEP (1981c) *Connect* Vol. VI, No. 3. September 1981.

UNEP (1987) *Connect* Vol. XII, No. 3. September 1987.

UNEP (1988) *Connect* Vol. XIII, No. 2. June 1988.

UNEP (1992) *Connect* Vol. XVII, No. 2. June 1992.

UNCED (United Nations Conference on Environment and Development) (1992). *The United Nations Conference on Environment and Development: A Guide to Agenda 21*. UN Publications Office, Geneva, Switzerland.

UNESCO-UNEP (1988a) *Environmental Education: a Process for Pre-Service Teacher Training Curriculum Development*. Environmental Education Series 26. UNESCO, Paris.

UNESCO-UNEP (1988b) *International Strategy for Action in the Field of Environmental Education and Training for the 1990s*. UNESCO - UNEP, Paris / Nairobi.

Vinke, J. (1992) 'Actors and approaches in environmental education in developing countries'. In Schneider, H. (Ed) *Environmental Education: An Approach To Sustainable Development*. Organisation for Economic Co-operation and Development (OECD), Paris, France.

Chapter 2

THE INTERNATIONAL ENVIRONMENTAL EDUCATION PROGRAMME'S CONTRIBUTION TO WORLDWIDE ENVIRONMENTAL EDUCATION

Wimala Ponniah
UNEP
UN Building
Rajdamnam Avenue
Bangkok 10200
Thailand

INTRODUCTION

Only in the last 25 years or so has there been a growing world concern for the future of mankind in the face of a rapidly deteriorating environment. The reason for this widespread concern may be attributed to the fact that most environmental matters hold global significance. Resulting from this concern, much attention has been focused on the effects of pollution, the exponential growth of populations, the deterioration of the Earth's natural resources (*viz.* forests, wildlife, fishing grounds, fertile soil) resulting from spiraling demands for energy and consumer products and more recently on global warming, the depletion of the ozone layer, biological diversity and on the international transport and disposal of hazardous wastes.

Global concern about the world's environmental problems - development and its environmental impact, the endangered quality of human life - led to the unprecedented United Nations Conference on the Human Environment in June 1972 in Stockholm. Specifically, 'Recommendation 96' of the Stockholm Conference called on "the organisations of the UN System, especially UNESCO... (to) take the necessary steps to establish an international programme in environmental education, interdisciplinary in approach, in-school and out-of-school, encompassing all levels of education and directed towards the general public."

FROM STOCKHOLM TO BELGRADE AND BEYOND

In response to 'Recommendation 96' of the Stockholm Conference, UNESCO and UNEP - the latter itself established by a recommendation of the Stockholm Conference - jointly created the current 'International Environmental Education Programme' (IEEP) in January 1975. The IEEP may be considered as a co-operative response at the international level to the pressing concern of nations about their threatened or already damaged environments, thus laying the foundation for a world-wide environmental education programme that will make it possible to develop new knowledge and skills, values and attitudes, in a drive towards a higher quality of life for present and future generations living within that environment. UNEP's direct involvement of the IEEP is handled by the Environmental Education and Training Unit (EETU) which works in consultation with the staff of other Programme areas. In addition, the Unit is responsible for directing implementation of decisions of the UNEP Governing Council on environmental education and training respectively. The IEEP was designed to:

a) facilitate the co-ordination, joint planning and pre-planning of activities essential to the development of an international programme in environmental education;

b) promote the international exchange of ideas and information pertaining to environmental education;

c) co-ordinate research to understand better the various phenomena involved in teaching and learning;

d) formulate and assess new methods, materials and programmes (both in-school and out-of-school, youth and adult) in environmental education;

e) train and retrain personnel adequately to staff environmental education programmes; and

f) provide advisory services to Member States relating to environmental education.

It appears that, from the beginning, the accent of the IEEP has been squarely placed on service to, and participation of, UNESCO's Member States.

In October 1975, this Programme held an International Environmental Education Workshop in Belgrade, in the former Yugoslavia. This workshop grappled with the view that different nations can no longer isolate themselves in their 'ivory towers' and keep rejecting worlds outside their own. Development and conservation were both to be considered as within the bounds of environmental education. The delegates attending the International Environmental Education Workshop drew up and unanimously adopted the

'Belgrade Charter' which espoused a 'new global ethic' and called for "changes which will be directed towards an equitable distribution of the world's resources and will more fairly satisfy the needs of all peoples." This charter or global framework provided an excellent frame of reference for the task of designing environmental education programmes from the realities of the situation. It recognises the urgent need in environmental education to develop a global understanding or perspective of the ecological, economic and moral considerations involved.

As a follow-up to the Belgrade Workshop, UNESCO and UNEP gave their support to a series of innovative pilot projects throughout the world selected in accordance with the guidelines and recommendations drawn up at Belgrade. As another part of the follow-up to the Belgrade Workshop, regional meetings were held during 1976 and early 1977 in Africa, the Arab States, Asia, Europe, Latin America and North America, which brought together representatives from all over each region to review and evaluate the Belgrade recommendations in more specific regional contexts prior to a World Conference at governmental level which was convened in Tbilisi, Georgia in 1977.

The Tbilisi Conference marked the culmination of the initial phase of the IEEP and set the stage for intra- and international - efforts which were to follow during the next few years. The Tbilisi Conference also confirmed - and stressed the re-enforcement of - aims and priorities of the IEEP. The recommendations from the Tbilisi meeting emphasised 'quality of life' as opposed to 'environmental quality' alone. Thus a more human-focused definition of environmental education was accepted as the foundation for the global efforts which were to come.

In August 1987, ten years after Tbilisi, the IEEP convened the UNESCO-UNEP International Congress on Environmental Education and Training in Moscow, Russia. This Congress was aimed at high-level administrators and decision-makers. In his opening address, the African Secretary-General of UNESCO emphasised once again the role of education in meeting environmental challenges:

"This is why one of the most important items on the provisional agenda before you concerns the role of education in facing the challenges of environmental problems. In an area as new, after all, as that of environmental education, clarification of concepts and principles which should guide our action is of crucial importance. What must be done is to state as clearly as possible not only what is meant by environmental education, but also, indeed above all, its specific functions as part of the general effort to bring about a renewal in education in order to prepare   each individual squarely to shoulder his responsibilities."

Out of this activity emerged the 'International Strategy for Action in the field of Environmental Education and Training for the 1990s', which has served as a major blueprint for governments and local and international organisations in the orientation of their environmental education and training policies and actions.

To date the Strategy has been widely disseminated in English, French, Spanish, Russian, Arabic, Chinese, Japanese, German and Hindi, thereby providing member states and institutions with a framework and guidelines for the present decade and a basis for preparing their own national strategies for environmental education and environmental training for the 1990s. During the last few years, efforts were jointly made by UNESCO and UNEP to encourage national governments worldwide to each prepare a 'National Environmental Education and Training Strategy for the 1990s' based on the above-mentioned International Strategy.

FIELDS OF ACTION OF THE INTERNATIONAL ENVIRONMENTAL EDUCATION PROGRAMME AND ITS ACCOMPLISHMENTS

In accordance with needs and priorities of Member States, the IEEP has envisaged a progressive strategy involving different levels and modalities of education (primary and secondary school education, non-formal education). A multiple actions have been undertaken in this respect to develop and reinforce the principal elements of educational processes (information, pedagogical research and experimentation on content and methods, training of personnel, preparation of educational materials) and to provide Member States with appropriate advice for the incorporation of environmental education into their educational policies and programmes.

Outputs of the IEEP could therefore be properly appraised in terms of these principal areas: exchange of information and experience; research and experimentation; training of personnel; preparation of educational materials and regional and international co-operation.

Exchange of information and experience

This action is aimed at providing Member States and non-governmental institutions, as well as specialists and the general public, with information and opportunities for the exchange of experiences relating to research, policies, programmes, activities, training of personnel, materials and publications in the field of general environmental education. The accomplishments include:

- The first Intergovernmental Conference on Environmental Education (Tbilisi, 1977);
- The International Congress on Environmental Education and Training (Moscow, 1987);
- The newsletter 'Connect', distributed quarterly in 8 languages (Arabic, Chinese, English, French, Hindi, Russian, Spanish, Ukrainian) to over 200,000 readers;
- Copies of IEEP documents distributed annually;
- 'International Directory of Institutions Active in the Field of Environmental Education', distributed in English, French and Spanish;
- 'International Strategy for Action in the Field of Environmental Education and Training for the 1990s', distributed in 8 languages (Arabic, Chinese, English, French, German, Japanese, Russian and Spanish).

Research and experimentation

This action is aimed at promoting and developing research activities dealing with content, methods, organisational mechanisms and training strategies for furthering formal and non-formal environmental education. It is also devoted to the promotion of experimental and pilot activities stimulating environmental education practice in Member States. The accomplishments include:

- more than 138 pilot, experimental and research projects and studies to help Member States incorporate environmental education in their educational policies and plans, school curricula, teacher education, university teaching, and non-formal education.

Training of personnel in environmental education

The Tbilisi Conference stated that "the training of qualified personnel was to be a priority activity". This holds good for both pre- and in-service training, for the purpose of familiarising teachers in formal education, organisers in non-formal activities for young people and adults, administrative personnel and educational planners and researchers with environment-linked subject matter and educational and methodological guidelines. It was further emphasised that environmental sciences and environmental education should be included in curricula for pre-service teacher education, that Member States should take the necessary steps to make in-service training of teachers in environmental education available to all who need it, and that education and training institutions should have the necessary flexibility to enable them to include appropriate aspects of environmental education within existing curricula and to create new environmental curricula which meet the requirements of an interdisciplinary approach and methodology. The accomplishments include:

- 45 global, regional and sub-regional training seminars and workshops throughout the world, and 140 national training workshops for orientation of key educational personnel in Member States.

Educational materials and publications

This action involves the preparation, adaptation, production and diffusion of educational materials (manuals, guidebooks, modules, audio-visuals, etc.) as well as publications concerning various levels and forms of environmental education and training. The accomplishments include:

- more than 54 environmental education curriculum prototypes for primary and secondary schools and guidelines for environmental education development, environmental education modules on environmental themes; 18 environmental education basic documents and 49 reports.

Regional and international co-operation

This action involves advisory missions to Member States for the purpose of developing curricula and setting up training activities relating to environmental education. Technical support continues to be given to national educational institutions, governmental and non-governmental organisations for the promotion and development of this area of their activities.

IMPACT OF THE ACTIONS OF THE IEEP

Contributing most to the development of global awareness of environmental education needs was IEEP's series of international and regional meetings culminating in the Tbilisi Conference and the Moscow Congress. A policy of regular, periodical information also contributed - and continues to contribute - to this awareness, notably through the IEEP's international newsletter *'Connect'*, published in Arabic, Chinese, English, French, Russian and Spanish, and distributed freely each quarter to 200,000 individuals and institutions throughout the world who are actively involved in environmental concerns, as well as in environmental education. In parallel, the IEEP has been steadily building a computerised information system whose data base contains information on over 1,500 environmental education institutions and projects, which is published in regularly updated directories.

The International Environmental Education Programme itself has undertaken dozens of pilot, experimental and research projects aimed at aiding nations to incorporate environmental education into their educational processes. Similarly, IEEP has organised many regional and sub-regional environmental education teacher-training workshops in all regions of the world as well as national training workshops in the same regions. As a result, about 125 countries have been directly involved in IEEP's activities which include preparation and diffusion of environmental education manuals, guidebooks, modules, sourcebooks, audio-visual and other materials concerning all levels and forms of environmental education and training. Of particular interest to those involved in basic education is the 'EE Series' of publications of IEEP, especially the teaching and learning modules for primary and secondary education.

One might also note the catalytic *multiplying effect* of the IEEP, the environmental sensitisation and training of thousands of key personnel, among them teacher educators, who have in turn trained many more teachers, and others, who have raised the environmental awareness and sensitivity of their countries' citizens.

Thus, through its activities, the IEEP has directly involved more than 150 countries from all regions of the world, more that 250,000 pupils in primary and secondary schools, about 10,000 teachers, educators and educational administrators and has contributed to environmentally sensitising untold thousands of persons.

ENVIRONMENTAL EDUCATION AT UNIVERSITY LEVEL

Realising that environmental education has an important part to play in general university education as well as in the training of specialists whose professional activities after graduation are likely to have an impact on the environment and its associated problems, their prevention and solution, the IEEP has, since 1984, been giving new emphasis to environmental education at the university level, particularly in general university education. The activities include a survey as well as studies of the current situation, regional seminars, experimental projects and training workshops. This aspect of the Programme was conducted in the light of recommendations and conclusions of the Tbilisi Intergovernmental Environmental Education Conference of 1977. Recommendation 13 specifically states that "the Conference, considering ... that environmental education in colleges and universities will become increasingly different from traditional education and will teach students basic knowledge for work in their future profession, which will benefit their environment, recommends to Member States (*inter alia*):

- To encourage acceptance of the fact that, besides subject-oriented environmental education, inter-disciplinary treatment of the basic problems of the interrelationships between people and their environment is necessary for students in all fields, not only natural and technical sciences but also social sciences and arts, because the relationships between nature, technology and society mark and determine the development of a society;

- To develop different teaching aids and text-books on the theoretical bases of environmental protection for all special fields to be written by leading scientists as soon as possible;

- To develop close co-operation between different university institutions (departments, faculties, etc.) with the specific objective of training experts in environmental education;

- Such co-operation might assume different forms in line with the structure of university education in each country, but should combine contributions from physics, chemistry, biology, ecology, geography, socio-economic studies, ethics, education sciences, and aesthetic education, etc."

Two special groups of university students are mentioned as target audiences for more intensive environmental education and training following that given to students in general. They are: student scientists, technologists and other future experts and professionals who will be dealing directly with environmental concerns (foresters, biologists, hydrologists, ecologists, agriculturalists, and the like); and those students of specific professions and social activities whose future work will have an influence and impact on environmental management, both rural and urban, somewhat less directly (engineers, architects, urbanists, economists, labour leaders, industrialists, *etc.*).

The IEEP recognises that the most immediate and practical approach to the introduction of environmental education to university students on a multi-disciplinary basis has been to incorporate environmental themes - an environmental dimension - into most, if not all, existing disciplines of the various faculties of the university, most obviously in such disciplines as the natural and human sciences, but not limited to them. If the local or regional environment and its problems are the point of reference in each case, then, in effect, there would be a kind of co-ordinated team-teaching. Incorporation of the students' environment, more over, would give each discipline the relevance that is now so much sought, making it alive and pertinent, rather than dryly academic - particularly if participation in problem-solving is part of the environmental education process. In this sense, too, the university would be educating for responsible citizenship, a goal high on, if not at the top of, its educational priorities.

Also, the IEEP recognises that the most revolutionary approach would be not simply to add an environmental dimension or even component to a discipline - such as creating new disciplines of environmental chemistry, environmental law, environmental economics, etc. - but to conceive the whole discipline as revolving around the environment, viewed as a complex whole. This involves the conception of one's environment in its totality - natural and built, taking into account air, water and soil, urbanism and land management, economics and esthetics, sociology and ethics, history and technology, *and* politics. This is the essence of the renovation in education which a comprehensive environmental education brings about.

ENVIRONMENTAL EDUCATION FOR TECHNICAL AND VOCATIONAL STUDENTS

After giving new emphasis to environmental education at the university level, particularly in general university education, the IEEP in 1987 placed visible accent on the promotion and development of environmental education in technical and vocational education.

In this respect, the IEEP recognised that engineers and agronomists, industrial and agricultural technicians, specialised and skilled educators, workers, that is, the principal graduates of systems of technical and vocational education, need special environmental education for two reasons: they participate in development and production processes which have the greatest impact on the environment and they themselves are often most affected by that impacted environment as well as by the particular hazards and accidents of their work place

33

or 'inner environment'. Environmental education has taken both roles into consideration. In addition, it encompasses both initial and on-the-job training in its scope.

The IEEP also recognises that technical staff at all levels have a primary responsibility for the protection of the environment and for improving it whenever possible. They are directly engaged in the development of new products and materials, and responsible for the design of work processes, for the development, operation and control of technical installations, for ensuring that process outputs and by-products are properly designed in line with established standards and that these standards are properly maintained, and that wastes are properly disposed of.

Thus environmental education in technical and vocational education, as advocated by the IEEP, aims at making students and trainees aware of the specific environmental problems and risks, including those relating to the safety and health of the working group, that relate to the type of industry and occupation which the students are preparing to enter, and which are involved in the type of functions that they are learning to undertake in economic life. In addition, they learn the theoretical and scientific basis of the technology for coping with environmental problems and acquire the skills needed for applying it. The training relates to the construction, design, use and maintenance of protective measures and to the technology that might be applied in reducing or removing the relevant environmental risks.

THE IEEP'S CONTRIBUTION TO UNCED

During the June 1992 Earth Summit in Rio de Janeiro, Brazil, there was strong support expressed for sound environmental management and sustainable development from the 30,000 participants and some 120 Heads of Government or State who attended the Conference.

Most of what was agreed in Brazil by way of 'Agenda 21' and the Rio Declaration was not new to the environmental community. In fact, roughly 95 per cent of Agenda 21, which deals with a broad range of issues from forestry management to the protection of freshwater resources and from science for sustainable development to atmospheric protection, had been defined in other separate fora and documents. Yet, what is new about Agenda 21 is that it brought most of the major environment and development issues under one roof.

Chapter 36 of Agenda 21, entitled 'Promoting Education, Public Awareness and Training' represents the contributions from the Environmental

Education Sections of UNESCO and UNEP through a working group constituted specifically for that purpose. The programme areas of Chapter 36 include:

- Re-orienting education towards sustainable development;
- Increasing public awareness;
- Promoting training.

The IEEP in the post-UNCED process will be guided by the following set of goals of environmental education for sustainable development:

- sensitising individuals and groups, communities and nations to ecological, economic, social and cultural interdependence;
- providing everyone with the opportunity to acquire awareness, knowledge, skills and commitment in order to protect and improve the environment;
- creating new environment-friendly behaviour patterns;
- developing environmental ethics;
- fostering environmental literacy for all;
- improving the quality of life.

The first two of the above-mentioned goals are not new as they have been enshrined in past phases of the IEEP. They have been repeated here because of their important sensitising and awareness-building roles *vis-à-vis* environment and development issues. The remaining four are intended to re-shape education towards sustainable development.

A major priority of the UNCED recommendations is to re-orient education towards sustainable development by improving each country's capacity to address environment and development in its educational programmes, particularly in basic learning. This is seen as indispensable for enabling people to adapt to a swiftly changing world and to develop an ethical awareness consistent with the sustainable use of natural resources. Education, in all disciplines, is expected to address the dynamics of the physical-biological and socio-economic environment and human development, including spiritual development, employing both formal and non-formal methods of communication.

The Earth Summit's 'Agenda for Change', written by Michael Keating and published by the Centre for Our Common Future, explains what nations should seek to do to improve sustainable development education:

- Make environment and development education available to people of all ages;

35

-     Work environment and development concepts, including those of
      population, into all educational programmes, with analyses of the causes
      of the major issues. There should be a special emphasis on training
      decision makers;

-     Involve schoolchildren in local and regional studies on environmental
      health, including safe drinking water, sanitation, food and the
      environmental and economic impacts of resource use.

It is most encouraging that UNCED has assigned to IEEP the task of assisting in
the development of an integral, coherent educational approach across the broad
areas of environmental protection, sustainable development, science and
technology knowledge (Agenda 21, Chapter 36.5):

"Within two years the United Nations system should undertake a comprehensive
review of its educational programmes, encompassing training and public
awareness, to reassess priorities and reallocate resources. The UNESCO / UNEP
International Environmental Education Programme should, in co-operation with
the appropriate bodies of the United Nations system, governments, non-
governmental organisations and others, establish a programme within two years
to integrate the decisions   of the Conference into the existing United Nations
framework adapted to the needs of educators at different levels and
circumstances."   (Agenda 21, Chap.36.5(g), UNCED, Rio, 1992)

In view of the above recommendation, the Director General of UNESCO
convened a Consultation Meeting among UN organisations including UNEP and
some intergovernmental and non-governmental organisations concerned with
environmental education and information at the UNESCO Headquarters in Paris
in September 1993.
      The aim of the meeting was to foster exchange of information within the
UN system and inter-governmental and non-governmental organisations
concerned with environmental education and discuss ways and means of making
an in-depth review of policies and methods regarding the promotion of
environmental education, public awareness and training - the ultimate objective
being the establishment of a coordinated UN interagency programme integrating
the UNCED decisions related to environmental education and information.
      A number of concrete proposals were made during the interesting
discussion that followed and a certain number of clarifications - regarding the
precise means and manner of use regarding exchange of information; the

parameters to be used in the evaluation of activities; co-operation and co-ordination at the country level etc., had to be made. At the end of the discussion, the Consultation Meeting came to the following major decisions:

- Each organisation / agency should develop its respective environmental education data base for possible inter-connection and designate active focal points for expediting information exchange on environmental education activities as a follow-up to UNCED.
- UNESCO should prepare, on the basis of self-evaluation inputs from the UN bodies covering a period of at least two years, a UN system-wide comprehensive review of environmental education and submit it to the Commission on Sustainable Development.
- Each UN body should send the environmental education component of its draft programme and budget for 1994-1995 and its next draft medium-term plan to UNESCO for the preparation of a joint programme to integrate UNCED decisions related to environmental education into the existing United Nations framework.
- Environmental education collaboration at country level through the development of coherent programmes and involving concerned UN bodies and national organisations should be fostered.
- Development of national strategies and action plans in environmental education as an integral part of national plans for education, health, agriculture, industry and socio-cultural development should be actively encouraged.
- Establishment of national co-ordinating boards in the field of environmental education should be promoted in order to facilitate decision-making within the countries and with various donors, especially the UN bodies.
- Greater co-operation with non-governmental organisations, the private sector and the media should be sought in order to ensure an effective impact on all sections of the population.
- Strong efforts should be made to identify funding sources outside the UN system for faster and wider implementation of activities mentioned in Chapter 36 of Agenda 21.

CONCLUSION

As mentioned earlier, the IEEP is a co-operative response at the global level to the urgent concern of nations about their threatened or already damaged environments. It has been, perhaps, the most successful inter-agency collaborative venture between UNEP and UNESCO. The IEEP, which is 20 years old, has no closing date as the joint partners UNESCO and UNEP believe that there should be no end to efforts to preserve and improve the environment for generations to come. Never has the impact of societies and their development been deeper and wider than on their environments.

The challenge for the IEEP in the post-UNCED era is clear - to devise long-term environmental strategies for achieving sustainable development and to reach out to new audiences, including economists and ecologists, technicians of most kinds, sanitation workers, researchers, planners and designers, architects and engineers, farmers and foresters, fishing folk, people in industry and trade, other 'grass-roots' decision-makers and mass media specialists. Since 1975, IEEP has:

- contributed to the world's general awareness about, and clarification of the concept of the environment as encompassing both natural and built components;
- fostered through environmental education the recognition, solution and prevention of environmental and development problems at global, regional and national levels;
- contributed to the identification of needs and priorities, to the clarification and development of the philosophy, objectives and guiding principles of environmental education that have served as a common denominator in the educational curricula renewal of Member States;
- developed environmental education guidelines and strategies as well as curriculum prototypes, and promoted their local adaptation;
- trained key educational personnel serving as a 'multiplier effect' on the development of environmental education - especially its incorporation within the curricula for primary and secondary schools, technical and vocational education, teacher training, and higher education;
- fostered international co-operation in environmental education through technical and educational support, field missions, and participation in relevant activities of international governmental and non-governmental organisations;
- provided inputs to UNCED and consequently Chapter 36 of Agenda 21 based on the fundamental principles of the Declaration and Recommendations of the Tbilisi Conference on Environmental Education organised by UNESCO and UNEP and held in 1977 (Para 36.1 Agenda 21).

As a consequence more than 95 countries have adopted environmental education as a key component of national formal and non-formal education. Curriculum prototypes for primary and secondary schools and for teachers - prepared on the basis of sub-regional environmental and educational needs and priorities for Africa, the Arab States, Latin America and the Caribbean and the Asia-Pacific region - will serve, through local adaptation, as seeds of reorienting education towards sustainable development in about 125 countries having a total population of 4.5 billion people.

The challenge of Member States to, and the expected response of the UNESCO-UNEP IEEP, is a dynamic continuum of more encompassing environmental education activities among more and more Member States, for there is a consensus that *all* are involved in the environment of *each* and *each* in the environment of *all* - the world environment.

Chapter 3

CARING FOR THE EARTH: CASE STUDIES IN ENVIRONMENTAL EDUCATION AND COMMUNICATION

Frits Hesselink
SME
PO Box 13030
3507 LA Utrecht
Netherlands

INTRODUCTION: LEARNING FOR SUSTAINABILITY

Integrating the 'what' and the 'how' of sustainable development is the essence of environmental education and communication. It is a double-edged challenge. For all our knowledge it is not clear what equity and sustainable living will mean for all Earth's inhabitants a generation from now. The process is a learning experience for the whole of society, not just in schools or among young children. We must learn to develop the how - the processes and modes of enquiry and learning - so that we are better at motivating, informing and supporting all the sectors of society on the path to sustainable development. Governments, social groups - both business and organisations - and, of course, local communities, have a crucial part to play. This chapter shows how some of IUCN's network of members, programmes and the Commission on Education and Communication (CEC) are helping in this process.

WHAT CONSERVATIONISTS SHOULD KNOW

Despite the accumulated knowledge of conservationists and all our efforts to implement the principles of sustainability, we are still not making great progress. Why is this?

My answer is: because conservation is a matter of people, and people need to know what to change in their lives and how. This is the challenge for educators and communicators. Conservationists must learn to use the tools and skills of communication, just as educators must be aware of conservation principles and practices.

But are our citizens willing to put conservationists' ideas into everyday practice, in the office, in the home, in their free time? Since most societies are not using natural resources sustainably or equitably, this will mean changing behaviour. And such a restructuring of society is best driven by a change in ethic and attitudes.

We, the education and conservation professionals, need to know 'what' has to change as well as 'how' that change should take place. What internal motivation will contribute to making it happen? Governments, for example, must learn how to be more informed of grassroots activities, and how to maintain a dialogue with these organisations, and to communicate to others the successes.

Environmental messages are more credible when they come directly from a source close to the people being addressed. Workers, for example, are more likely to take messages more seriously when they come from the unions, than when they come from the State. The same goes for business executives: for them, chambers of commerce are more reliable sources than governments.

But governments can increase the impact and potential of grassroots success by providing information and subsidising activities, materials and facilities. The rewards are increased credibility, effectiveness and quality of results.

Many governments are providing frameworks for these co-operative ventures as our case studies from Australia, Canada, Nepal and the Netherlands (edited by Peter Hulm) show in this chapter. These efforts provide encouragement, marketing, information, better communications between groups, partnerships with grassroot organisations and other government sectors.

The work by members of IUCN and the CEC network featured in the following pages is enabling people to learn for themselves what needs to be done, and motivating them to change and adapt their lifestyles. We do not have all the answers, but the experience we have accumulated can provide guidance for IUCN and other organisations working to bring the message of 'Caring for the Earth' into the homes and communities of people everywhere.

Major contributors to this chapter included Kathleen A. Blanchard, Peter Bos, Andrew Campbell, Juliana Chileshe, Marco A. Encalada, Jaime González Acosata, T. Christine Hogan, Stella Jafri, Chris Mobbs, Suzana M. Padua, Badri Dev Pande, Jim Taylor, and Paul Vare.

NETHERLANDS: ENVIRONMENTAL SUPERMARKET

In five years, the Government of the Netherlands (a State Member of IUCN) has nearly tripled its spending on environmental education to some US$18.2 million a year. How did this come about? Seven Ministries are combining their resources in support of a national strategy in both the formal and non-formal educational sectors.

This new initiative results partly from a 1988 White Paper on Environmental Education, leading to calls by Parliament for Ministerial co-ordination. A major breakthrough came in 1990 with the decision, by several government departments, to make a joint effort in the formal education system. New programmes support - rather than replace - existing governmental education activities. About US$40 million has been allocated for 1992-95. Projects have been taking place in vocational and agricultural education institutions as well as in primary and secondary schools.

To develop a strategic plan for environmental education focusing on the non-formal sector, the government called in a consultant in 1992 and received advice from IUCN's Commission on Education and Communication. This resulted in a General Plan on Environmental Education presented by the Dutch Cabinet late in 1993.

The plan focuses strongly on incorporating environmental education into the daily business of all sorts of groups and organisations. NGOs and government agencies have developed and supplied teaching kits, handbooks, courses, excursions, training, fieldwork and the like. In all, seven Ministries, several dozen NGOs, 12 provinces and over 700 local councils are involved in this new effort. An independent manager has been commissioned to map out operational fields of action.

"Environmental education," says Dr Peter Bos, chair of the Interdepartmental Working Group for Environmental Education in the Netherlands, "has grown into a sort of supermarket, an enormous superstore that covers a vast number of products, clients and suppliers."

More and more organisations are already acknowledging their co-responsibility to initiate or co-sponsor activities aimed at sustainable development. The Dutch housewives' organisation (Nederlandse Vereniging van Huisvrouwen), together with 12 other organisations, is taking part in a campaign to reduce car use. Community work has begun on local projects to care for nature and the environment. A group of development organisations has made a commitment to environment and development activities as a follow-up to the 1992 United Nations Conference on the issue.

CANADA: CREATING ENVIRONMENTAL CITIZENS

When Canada (an IUCN State Member) was developing its Green Plan, it found out, through a survey performed in 1992, that 84 per cent of its citizens believed that their behaviour had to change if the environment was to be protected. The survey also revealed that they were prepared to make those changes voluntarily, if they had credible information about what to do.

Canada's Environment Citizenship Programme, launched in mid-1992, aims to create a culture for voluntary action by an informed and educated citizenry. A major strategy of the Government has been to seek partnerships with other organisations.

Environment Canada, a government department, has developed information materials such as primers on fresh water, ozone, waste management, spaces and species and global warming in collaboration with organisations like Friends of the Earth, DuPont Canada and Parks Canada. These primers are used in Environment Canada's own programmes and in those of other bodies such as the Association of Community Colleges of Canada, which is making them a basis for education programmes in 700 colleges. Businesses, key providers of education to adults, are also adopting environmental education programmes as well as Municipalities.

One innovative way in which Environment Canada is providing information is through its meteorological office. Since February 1993 environmental citizenship messages have been disseminated through its network of weather offices to more than 250 media outlets and have become a regular part of daily weather forecasts. The media access the messages by contacting their local weather office or through subscriptions to national and regional news agency services.

ZAMBIA: LISTENING TO PEOPLE

The 1985 National Conservation Strategy (NCS) of Zambia (a State Member of IUCN), said that a comprehensive programme for conservation education was "the surest long-term strategy to bring about the sustainable use of natural resources in Zambia". However, the NCS was prepared without broad-based participation and many Zambians are unaware of its existence.

As part of Zambia's Environmental Education Programme (ZEEP) a survey was taken to find out beforehand what people knew and thought about the environment. Some of the views expressed:

43

- "They asked us to move from our original village so that they could create a national park. Who benefits from the national park?"
- "Rich men with big guns come to hunt while we look on. They kill large numbers of animals while we starve for meat."
- "One can never have enough children."
- "It is not my duty to clean up."
- "Fish drops from the sky with the rain so it cannot vanish from the rivers."
- "If you advise to plant trees, when shall we plant maize?"
- "Nobody plants trees. They grow on their own."

Co-ordinator Juliana Chileshe, Deputy Chair of CEC, underlines: "ZEEP has tried to establish what people say, do, and think but do not express about the environment." Her conclusion: "In order to do away with misconceptions and to change popular attitudes - often negative to environmental protection - the benefits of conservation have to be apparent."

Zambia's recently introduced National Environmental Action Plan, which the WWF International-backed education programme supports, aims to implement the National Conservation Strategy through such means. It will develop projects that encourage people to use natural resources - through fish farming, game ranching and the creation of forest reserves - without exhausting the environment.

NEPAL'S NCS: EDUCATION IS THE KEY

Since the middle of 1989, IUCN has been helping implement Nepal's National Conservation Strategy, endorsed in 1988 after five years of consultations.

Environmental education is given an important role in putting the NCS of Nepal (an IUCN State Member) into effect. The Eighth Five-Year Plan (1992-1997) provides for environmental education at all levels of formal education, technical education, teacher training, non-formal adult education, and in-service programmes. The Plan also stresses the need to use the mass media to raise the level of environmental awareness.

"One of the first initiatives was to review courses, textbooks and teaching materials," says Dr Badri Dev Pande, Co-ordinator of the Environmental Education and Awareness Programme of the National Planning Commission / IUCN NCS Implementation Project in Nepal. "We found that all contained some element of environmental education, but they were still inadequate in inculcating attitudes and encouraging action that would prevent deterioration of the environment."

Over the past two years, after training workshops, an evaluation seminar and a national environmental education conference, resource materials in environmental education have been tested with nearly 2500 primary school pupils from 10 experimental and 10 control schools in Dhankuta, Kathmandu, and Dang. The trial has led to IUCN having a major influence on revising formal educational curricula to integrate environment and development.

Environmental camps and environmental art workshops and competitions are also being organised for primary and secondary schoolchildren, in collaboration with local NGOs. "Although these are able to reach only a limited number of students, the extra-curricular activities have achieved some notable results," Dr Pande observes. "A high school headmaster in Dang confessed that as a result of his son's taking part in an environmental camp, he decided to install a smokeless oven in his house."

Environmental education is incorporated into primary school teaching in four subjects: Nepali language, social studies, health education, and science. "Teachers have reported that the majority of children involved in the experimental programme found the texts so interesting they would read them before they reached the lesson concerned," says Dr Pande. "Some children even planted flowerbeds and kitchen gardens of their own as a result, and many improved their sanitary habits."

"The challenge now is to maintain the level of enthusiasm and motivation currently found in students, teachers and trainers taking part in the programme," Dr Pande remarks. "We have learned that partnership with NGOs is valuable in securing timely technical assistance. One NGO helped prepare illustrations for the primers, and a journalists' forum puts together and distributes a wall newspaper for remote villages."

Traditional means of communication, such as street theatre, can be used as a means to raise environmental and development issues among villagers and stimulate discussions with groups about solutions. The challenge for communicators is to find credible means to deliver their messages.

AUSTRALIA: A STRATEGY FOR THE RURAL COMMUNITY

The cost of European settlement has been high in Australia (a State Member of IUCN). In a little over 200 years, about 100 species of plants and at least 27 species of birds and mammals have disappeared, mainly as a result of human settlement. A further 209 plant species and 59 species of vertebrates are at risk.

Agitation about soil erosion, salinity and invasion of pests among rural communities, triggered both community and State government action in the

45

1970s and after. But 1988 saw a major development in co-operation to protect the environment: the National Farmers Federation and the Australian Conservation Foundation (an IUCN member) came together and proposed a land management programme. The next year this resulted in the government's 10-year National Landcare Programme, a A$250 million project for the 1990s providing for co-operative, co-ordinated approaches to solve problems at a district scale that cannot be effectively tackled at the level of individual properties. In spite of tough economic conditions facing rural communities, the Landcare movement has grown explosively. Today it includes over 2000 groups and involves about one third of rural Australian families.

"Landcare groups create collective social pressure in favour of more sustainable farming systems," says former Co-ordinator Andrew Campbell. "They reinforce and support the efforts of individuals, who are already working to reduce water quality decline, salinity and loss of diversity. They play an important role in gathering and managing information, education and awareness. They have created a community focus and networks of support. They share the stress of rural decline and promote action to do something about it."

Under the Landcare umbrella is the 'Save the Bush Programme' administered by the Australian Nature Conservation Agency (an IUCN member) and the 'One Billion Trees' programme, handled by Greening Australia. Co-ordinators and facilitators encourage and enable community action to re-establish and maintain Australia's native vegetation outside protected areas. Groups are given small grants, and information on plant identification, removal of weeds and exotic animals, planting of 'wildlife corridors' and are encouraged to fence areas from stock.

In this process, the role of the 'educator' has changed from teaching to facilitation and co-ordination. From the individual the focus has shifted to groups and involvement of a diversity of stakeholders. Above all there is the recognition that the complexities of tackling land degradation demand the involvement of those people whose daily decisions shape the land.

The role of the new 'educator-facilitator' is to develop a shared sense of direction among all the relevant actors, requiring insight into group processes. It is more a matter of skilled listening and of asking the right questions of the right people, at the right time, than of passing on technical information.

The government's role has been to provide funding to start group projects, co-ordinators and trained facilitators to stimulate the group process, communication about successful programmes and other community action, expert information on rehabilitation and identification of species, and marketing of the concept.

Despite progress the programme still faces constraints: limited human resources in rural areas, a lack of technically sound practical solutions to land degradation problems, institutional cultures in extension and agricultural research which work against genuinely participatory approaches, overwhelming technocratic training of professionals in extension and research, and a feeling among farmers that they are being blamed for degradation.

BRAZIL: BUILDING SUPPORT FOR PROTECTED AREAS

The biggest challenge for educators may be to reach people in time to safeguard valuable ecosystems, even in protected areas. Eighty-two per cent of the State of Sao Paulo, Brazil, was once covered by natural forest. Today less than five per cent remains. The last significant portion of interior Atlantic Forest is preserved as the Morro do Diabo (Devil's Hill) State Park in the far western part of that state. Deforestation threatens several animal species with extinction, among them the black lion tamarin (*Leontopithecus chrysopygus*), the most endangered species of primate on IUCN's Red Data List.

At the start of the 1970s, Morro do Diabo State Park was part of a 290,000 ha reserved area. Powerful landowners, political leaders and development projects have reduced the protected zone to only 34,000 ha. Even this area suffers from the impact of projects such as highways, railways, an airport and hydro-electric plants.

In an effort to raise awareness about the importance of conservation, an education programme was designed for local schoolchildren. But this proved too long-term for the immediate needs of conservation. "By the time students became decision-makers there might be little left to fight for," declares Suzana M. Padua, President of the Institute for Ecological Research (IPE) in Brazil and Director of the Brazil Programme for the Wildlife Preservation Trust International. Several out-reach programmes were thus initiated, targetting all sectors of the community, from local authorities to businessmen and labourers as well as schools.

By the end of the first year of the Black Lion Tamarin Environmental Education Programme, 6000 students had visited the Park and the average number in each of the succeeding three years was 8000. A before-and-after survey showed increases in knowledge and changes in attitude.

"There were other indicators of success," Padua notes. "Families visited the Park during weekends. The Park's lodging house began to be filled with university students looking for field experiences. Local teachers requested environmental education courses, and nature guides improved their performance by forming study groups. Community interest in the Park's conservation was demonstrated in the subjects chosen for floats in local processions for the town's anniversary, at year's end for Lions' and Rotary Club parties, and in public initiatives such as a tree planting campaign."

For Padua, however, the most important signs came from community involvement in protecting the Park. Garbage was being dumped in the vicinity of the Park, in an area which had been within its territory a few years earlier. The dangers were explained over the radio with an appeal to local leaders and the community to find a solution. The local population reacted by telephoning the radio station and demanding, from the Mayor, a solution. The garbage was removed in less than a week.

The community also helped Park employees to put out a forest fire. Although there had been many conflagrations in the past, the general public had never before been involved in fire-fighting.

"Local interest in nature conservation has also extended beyond the boundaries of the Park," Padua adds. "The community has spoken up against illegal logging taking place on a privately-owned farm nearby. Business entrepreneurs around the Park formed a group to plan development projects that were non-polluting and provided opportunities for underprivileged and unskilled local citizens."

UGANDA: PRACTISING PARTICIPATION

IUCN's Mount Elgon Conservation and Development Project in Uganda found itself forced into a 'top-down' approach for its education / extension programme. But to give more power to the local communities, the programme is practising "gradually diminishing control".

Mont Elgon's tropical montane forest is a vital water catchment area on the Kenya-Uganda border. But during Uganda's civil unrest in the 1970s and 1980s, over one-third of the forest was destroyed or degraded, chiefly as a result of agricultural encroachment.

The Conservation and Development Project, initiated in 1988, covers the National Park and the 58 neighbouring parishes - home to 223,000 people. The second phase (1991-December 1993) saw the creation of an Education / Extension Unit to work on training as well as awareness and support. Because of the short duration of the Unit's project, planning and implementation were almost simultaneous, and the project helped local organisers to run 23 practical seminars.

At the outset, the Unit had to create awareness about the project to protect the forest reserve. In this Project, top-down efforts seemed to fit local circumstances, says Paul Vare, the Unit's technical advisor at the time. "One cannot approach farmers without first gaining the approval of their local politicians. Even a consultative discussion is generally wrapped up with a short speech in which the farmers' representative gives thanks for the valuable education received. A people-centred, bottom-up approach can easily be misunderstood both by the experts and by the target group. Our sequence of awareness-raising events has been from district to sub-county level and thence to parish roadshows (drama performances and slideshows which regularly attract over 1000 people)."

As a result, the Unit applied a gradual approach to participation, stressing its importance in all seminars, and practising its technique at training courses. "Farmers' meetings are as much about gathering information as giving it and training courses are seen as opportunities to learn from the learners," says Vare. "Our fuelwood conservation courses, for example, include discussions with the female participants on how the project can reach women more effectively."

To generate longer-term commitment to conservation, the project has set up an exploration centre where young people from the region can stay overnight. For education programmes to assist communities manage resources, however, it became important to help people find alternative sources of incomes, for example through training in fish farming, better livestock management and bee-keeping.

SOUTH AFRICA: LOW-COST MATERIALS

Share-Net, set up by the South African Wildlife Society (an IUCN Member) in 1990, arose out of examining Wildlife Society experiences with workshops for both teachers and the public.

Activities were usually chosen because they worked well for the presenters. Informal talks with participants afterwards revealed that teachers seldom applied in a sustained manner the activities or concepts that had been demonstrated. As a result, the Society made the daily experiences of participant teachers and students an active part of the workshop discussions. These were: overcrowded classrooms, poorly qualified teachers and a lack of good quality, locally relevant educational materials.

So Share-Net, based in Howick, began an interactive process to develop low-cost, copyright-free resource materials. It had most success with a 'Hands-On' series of 10 booklets, providing simple reference material on inland and coastal ecosystems.

The booklets are sold in a simple shoebox-style display stand known as a cardboard library. Other products include simple black-and-white guides, water monitoring kits, activity packs, teachers handbooks and a resource pack on a local mining controversy. In response to demand, Share-Net later started a mail-order

49

service, and now the project almost pays for itself, thanks to contributions of time from environmental organisations and to use of the printing machines to carry out work for local businesses during quiet times.

SAHEL: THE LESSONS OF WALIA

Partly thanks to IUCN, environmental education in western Africa has become a priority of many organisations, leading at Ouhigouya, Burkina Faso, Sahel, to the creation of a large Environmental Education Sahel Network, operating in at least nine countries. IUCN's Walia approach involves publication of a magazine named after an emblematic local species, supplemented by visits from the publishing team to develop the themes and challenge students to speak up and to think about solutions. It now extends to four West African countries, from Mali where it began. The results can be seen not only in changed attitudes but also in action.

The IUCN programme is now run by locally trained staff. In schools the subjects highlighted in the magazine are brought to life through discussions and projects to solve local issues. This experience has taught teachers to produce cheap materials and develop new approaches.

CANADA: EDUCATION FOR BIODIVERSITY

On the North Shore of the Gulf of St. Lawrence in Quebec, Canada, seabirds have traditionally been harvested for food. But between 1955 and 1978 the populations of several species declined drastically because of illegal hunting and disturbance of breeding colonies. Within 10 years a programme developed by the Quebec-Labrador Foundation in collaboration with the Canadian Wildlife Service (CWS) helped not only to change attitudes among the 5600 people there, but also to implement effective conservation.

Most seabird populations breeding in sanctuaries along the Quebec north shore increased - some dramatically - between 1977 and 1988. The programme owed its success to many factors, said Dr Kathleen A. Blanchard, Executive Vice-President of the Quebec-Labrador Foundation (QLF), an IUCN Member. One was the long-standing reputation of QLF as a provider of social services on the coast. The small size of the population meant that new ideas could be quickly disseminated through well-established channels of communication. The influence of group leaders who endorsed the project was critical.

"The education programme promoted thoughtful, informed, positive behaviour (legal hunting, bird study) rather than focusing on stopping negative behaviour (poaching)" Blanchard observes. "The programme did not try to introduce activities that would be strange to the local culture (raising chickens, for example). Rather than attempting to convince people that it was wrong to kill birds, the programme acknowledged the cultural norm: that it is acceptable to harvest birds for an occasional meal."

CONCLUSIONS

The case studies described in this chapter illustrate a diversity of environmental education initiatives. The Commission on Education and Communication, one of IUCN's six specialised Commissions, is a network of professional educators and communicators who work in government, NGOs and educational institutions. The Commission seeks to encourage conservationists to recognise the value of using education and communication, as a way of motivating and engaging people in the task of conserving and managing the environment. By drawing on the expertise in the network, the Commission offers guidance on ways in which people can develop commitment to participate in caring for the Earth.

The Commission is particularly focusing its efforts on what governments can do to create a framework or strategy to stimulate environmental education and communication, as well as how NGOs can better plan their programmes for biodiversity conservation.

One of the challenges is for the conservation community to learn and reflect on the expertise of others both in the government sphere and at the grassroots. The case studies reported here are a glimpse of the successes and difficulties both the Commission and IUCN members have experienced in building commitment for the environment among communities. The Commission harnesses this experience as a basis for guiding further progress. If you would like to be part of this effort please contact us.

Chapter 4

ENVIRONMENTAL EDUCATION AT UNIVERSITY LEVEL IN A DEVELOPING COUNTRY CONTEXT: CURRENT TRENDS AND FUTURE CHALLENGES IN GUYANA

Paulette Bynoe

Environmental Studies Unit
University of Guyana, Turkeyen
Greater Georgetown
Guyana

INTRODUCTION

The effectiveness of environmental education as a complementary tool for environmental conservation has been widely recognised and advocated. According to Sterling (1992) "environmental education may have the potential to 'save the Earth'". The same idea was shared by Jarvis (1981) who expressed the view that education has a crucial role to play in the task of improving the management of the environment by imparting the knowledge, skills and determination necessary to resolve the problems that arise or are likely to arise. Environmental education is the cornerstone of long-term environment strategies for preventing problems, solving those which have occurred and assuring environmentally sound, sustainable development (Connect, 1992). In the context of developing countries, Boom (1994) views education as a driving force for political and economic transformation. According to the author:

"...education can play an important role in increasing awareness of environmental issues amongst the population of developing countries.... it will go a long way to encourage the right environmental policies to be formulated and appropriate actions taken in the field with regards to environmental protection and regeneration".

Moreover, Agenda 21 in referring to the subject of Public Awareness, Education and Training states:

"Education is critical for promoting Sustainable Development and for improving the capacity of the population to address and deal with environment and development issues" (UNCED, 1992).

ENVIRONMENTAL EDUCATION, POLICY AND LEGISLATION IN GUYANA

The importance of environmental education has long been recognised by the Constitution of Guyana. Indeed, it states that:

"Every citizen has a duty to participate in activities to improve the environment and protect the health of the nation." (Article 2:25) (Government of Guyana, 1980) and,

"In the interest of the present and future generations the State will protect and make rational use of its flora and fauna, and will take all necessary measures to conserve and improve the environment." (Article 2:36) (Government of Guyana, 1980).

In addition, the National Environmental Policy states that in order to conserve and improve the environment, the Government of Guyana will endeavour :

"to raise consciousness of the population on the environmental implications of economic and social activities through comprehensive education and public awareness programmes" and, "to involve the population, including indigenous peoples, women and youth, in the management of the environment and natural resources" (Government of Guyana, 1994a).

Furthermore, the Environmental Protection Bill of 1994 (Government of Guyana, 1994b) specifies one of the functions of the Environmental Protection Agency as "promoting public awareness of the ecological systems and their importance to the social and economic life of Guyana". In carrying out its functions the Agency will provide information and education to the public regarding the protection and improvement of the environment.

These developments form a background against which environmental education at university level in Guyana will be analysed. However, it may be useful at this time to provide an overview of the Guyana environment.

AN OVERVIEW OF THE ENVIRONMENT IN GUYANA

Guyana, located in the north eastern corner of South America, is the only English speaking country in the Southern American region. Its neighbouring countries are Surinam in the east, Venezuela in the west and Brazil in the south west. (Figure 1 illustrates the position of the country in the region). This tropical country has a total land area of 216,000 sq. km. Topographically, Guyana can be divided into four natural regions: the low coastal plain; the sandy rolling lands; the highland / Pakaraima region and the pre-Cambrian lowlands which include forested areas and tropical savannahs.

The economy

As is the case of most developing countries, Guyana's economy has not attained the development which its abundant natural resources could support. In fact, a significant degree of economic decline was experienced by Guyana in the 1970s and 1980s. This decline was partly due to weak public sector management combined with periodic weaknesses in the international markets for bauxite and sugar, which are two of the country's most crucial foreign exchange earners.

It is important to point out that in 1988 the Guyanese Government embarked upon a major change in its economic policy. This change involved the creation of opportunities for private sector investment in the economy as part of the Structural Adjustment Programme / Economic Recovery Programme. This strategy however has depended, to a large extent, on the promotion of growth in sectors such as agriculture, mining and forestry. The latter clearly indicates the need for a population which is well informed on environmental issues and the need for sound environmental management.

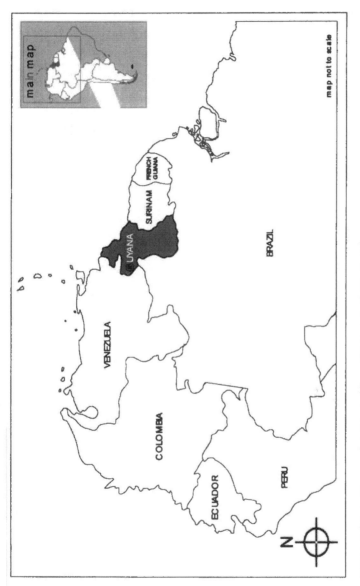

Figure 1 Schematic map showing position of Guyana in South America

Natural resource endowment

Guyana is endowed with an abundance of natural resources. These resources include the lush evergreen tropical rainforests which cover almost 80% of the country's total land area and contain over 1000 species of flora. The forests contain some plant species which are endemic to the country, such as the famous Greenheart and a rich variety of animal life, of which 144 are endangered species (Government of Guyana, 1992). In addition, the country has a substantial amount of mineral deposits including gold, diamond and bauxite, and very fertile lands on the low coastal plain.

Principal environmental problems

The major environmental problems of Guyana can be categorised as follows:

A) Coastal zone degradation: this includes coastal erosion, flooding and the impact of drainage and irrigation schemes.
B) Waste management and pollution control: particularly with reference to solid waste, agricultural chemical pollution, and environmental and health impacts.
C) Difficulties in implementing natural resources management, with regard to the management of watersheds, logging, forestry, wildlife and mining.

These environmental problems need urgent attention and should be dealt with in a fully integrated and comprehensive manner, as the environment is best managed in its totality. Failure to act in this way will probably have major effects on a country's development.

GENERAL TRENDS IN ENVIRONMENTAL EDUCATION IN GUYANA

Before attempting to understand the status of environmental education at university level, it may be useful to have an overview of the formal education sector in Guyana. Education has been free at all levels, with the exception of the university where fees were introduced during the 1994-1995 academic year. Literacy rates have been reasonably high (84%) and coverage from pre-school to secondary level has been broad. The education sector in Guyana, nevertheless, has been facing serious difficulties with reduced investments and inefficiencies in the use of resources. These reductions may have resulted in the deterioration of both the human and physical infrastructure and led to lower educational achievements. Very low salaries and poor working conditions have permeated *all* levels of the system, and have forced the better qualified teachers to leave the sector for more lucrative employment opportunities. This has led to a heavy reliance on untrained and unqualified teachers in the state education sector. According to a World Bank report in 1992, teaching aids and reference materials are lacking, and textbooks are insufficient or antiquated at all levels. The key constraint appears to be inadequate funding. In fact education spending as a percentage of GDP declined from 6% in 1984 to 2.4% in 1990, which is significantly below that of other Caribbean countries. It is interesting to note that Howell (1994) describes the formal education system as one of the most formidable Caribbean institutions. He adds that :

"It is guarded by strongly held values and complex bureaucratic arrangements which characterise the region, where the majority of teachers tend to resist any innovations which might even suggest an addition to an already demanding work load".

Obviously, this may have some implications for curriculum development for environmental education .

Generally, environmental education and public awareness programmes on environmental issues in Guyana have been conducted by a number of institutions, agencies and organisations in a sporadic manner, without any form of co-ordination or networking (Bynoe and Leal Filho, 1995). In some cases it has been developed as a reactive response to environmental crises, such as widespread flooding due to sea defence breaches, or pollution of rivers due to mining activities. In other cases programmes have been 'donor driven', so that input is drawn from outside the country with very little public participation and consultation.

ENVIRONMENTAL EDUCATION AT UNIVERSITY LEVEL

The need for higher education institutions to respond to the growing environmental crisis has been widely recognised, as well as advocated (Emmelin, 1975). No longer could universities in developed and developing nations turn a blind eye to their responsibilities as service institutions for society, or to the expertise demanded by labour markets or even the call for all human beings to become 'environmentally responsible citizens', each having a sense of responsibility for the environment.

"People who will have the greatest impact on the environment, for good or ill - tomorrow's policy and decision makers, engineers, architects, administrators in the public and private sectors, doctors, lawyers and teachers, not to omit environmental specialists are now...students at universities or higher education centres of learning. To impart to them the environmental knowledge and know-how for environmental care and improvement is literally a life or death matter for our planet, Earth... for all university students should have a high environmental literacy, forming an environmentally aware and concerned citizenry indispensable for all environmental decisions". (*Connect*, 1992)

"All students, whatever their course, make personal decisions which affect the environment; as such all students require environmental education for personal and social responsibilities".(*The Institution of Environmental Sciences*, 1993)

Moreover, the Belgrade Workshop in 1975 and the Tbilisi Conference held in 1977, recommended that environmental education in universities teach students basic knowledge for work in their future professions, which will benefit their environment.

It is reasonable to say that, prior to the above, the IUCN meeting on environmental education, held in Nevada (USA) in 1970, catalysed a number of efforts that have been made to promote environmental education at university level. Mention must be made to the Tours workshop held in France in 1971, which recommended that all universities should undertake environmental education at all levels, particularly within their normal curricula and courses. It was emphasised that universities should highlight the relation and contribution of each discipline and profession to the urgent problems of the environment and should encourage students to specialise in this aspect of their discipline or profession. Another significant meeting was the OECD sponsored conference on university level environmental education, held in 1972, which positively affected the events that followed in a number of developed countries, particularly Europe, America and Australia (Leal Filho, 1990). Today, it is widely acknowledged that universities in both developed and developing nations, have a crucial role to play in spreading, developing and applying knowledge among its students, to address the current environmental crisis, and more importantly, to promote sustainable development.

According to Agenda 21 (UNCED, 1992):

"To be effective, environment and development education should deal with the dynamics of both the physical/biological and socio-economic and human environment (which may include spiritual development), and should be integrated in all disciplines....."

Francis (1985) expresses the view that universities' responsibilities for delivering environmental education are:

a) to give instruction at undergraduate levels;
b) to pursue scholarship and research in environmentally related topics;
c) to offer informed comment and criticism of environmental issues, and
d) to assist with off-campus community initiated activities of various kinds.

Thus, universities have been encouraged to make the necessary provisions for environmental education for basically two groups of students: the specialists (scientists, technologists, foresters, biologists, ecologists, hydrologists and other future experts) and professions dealing with the environment. Secondly, and perhaps more importantly, the general university student whose discipline, whether it is Health Care, Law, Economics, Management, Architecture or Business Accounting, is part of the whole environmental problem and therefore needs to be taken into account in the process of finding solutions. This brings to mind, the term 'Greening the Curriculum' that has been introduced quite recently to place some kind of value on the environment.

GREENING THE CURRICULUM

Greening the curriculum is defined in a number of ways. A 'green' curriculum as defined by Ali Khan (1991) is a curriculum which helps students to understand the way their subjects of study relate to the environment, whether it is the natural environment, the man-made environment or the built environment. The greening of the curriculum is, however, more concerned with the integration of an environmental ethic into all courses. According to Irvine and Manns (1994) a meaningful 'green' curriculum goes beyond the teaching of 'hard' facts, (that is, from the description of environmental problems), to encouraging students to explore the roots of the environmental crisis, potential solutions and barriers to them. It would consider the present value systems and the broader changes necessary, behaviour patterns and institutional frameworks, as well as the limits to purely technological solutions.

The implication here is the whole interdisciplinary approach to environmental education. The ultimate aim is to enable all students to become environmentally responsible citizens. It is generally argued that it is not sufficient to green a part of the curriculum content, or more so the teaching methods, and ignore the vital lessons for life learned through the hidden curriculum of how the politics of the school are conducted. But, despite the rhetoric about environmental education at university level, there seems to be no evidence of widespread cases (of course there may be a few exceptions) of the full integration of environmental education in all university courses. In fact, the AUDES newsletter (Issue 1 March, 1995) in discussing the topic, notes that,

"Environmental education as an undergraduate experience is still to take its rightful place and be fully recognised as a legitimate and academically rigorous area of study, despite the support of many international, European, national and individual Universities, statements and declarations."

At the outset, it can be therefore be stated that, generally, environmental education at university level has not as yet received any significant status in the developing world in the context of the rest of the world (Leal Filho, 1990).

Approaches to curriculum development in environmental education

Ali Khan (1993) identifies three approaches to curriculum development for environmental education in universities. These approaches can be summarised as:

a) *Addition*: which assumes that the environment has little relevance to a particular discipline. Thus, the environment is dealt with in a manner that keeps it separate from what is perceived to be the learning agenda. The common practice is to establish a separate Department of Environmental Science / Studies.

b) *Incorporation*: which assumes that there is some relevance of the environment to the discipline. In this case, a number of environmentally related courses are introduced into traditional disciplines, using the prefix 'environmental' and

c) *Engagement*: which could change the shape of disciplinary enquiry, and may be associated with an action agenda for the environment. It has the potential for the achieving the goals of environmental education as discussed earlier in this chapter. This approach could be conceived as the most fundamental as it must be evolutionary.

It is against this background that the current initiatives of greening the curriculum, with specific reference to environmental education at the University of Guyana, needs to be seen.

Greening the curriculum at the university of Guyana

On a general note, there have been some very recent developments at the University of Guyana, with regard to environmental education teaching and research. These developments include :

a) The establishment of the Environmental Studies Unit in September 1993, in collaboration with the University of Utrecht in the Netherlands. The Unit will be involved in a number activities including: teaching about the environment; researching local environmental problems; providing expertise when necessary; promoting public awareness; undertaking consultancy work for private sector organisations; arranging public seminars.

b) The introduction of the undergraduate B.Sc. programme in Environmental Studies (the outline of this programme is provided in Appendix 1).

c) A new programme on Forest Biology has been developed at the postgraduate level and will be offered to students in the academic year 1995 - 1996.

d) A Chair for the "Sustainable Utilisation of Forests" has recently been established under the auspices of the United Nations Educational, Scientific and Cultural Organisation (UNESCO). The appointed Chair will be responsible for teaching and research and will be closely linked to the Commonwealth / Government of Guyana Iwokrama Rainforest Project. The fundamental objective of the latter is to demonstrate that the sustainable utilisation of forest resources is compatible with conservation of these same resources. To achieve this objective, a wilderness preserve at the project site and an International Centre for Research and Training for the Sustainable

Management of Tropical Rainforests, will be established. Additionally, the project will promote environmental literacy through formal and non-formal means (Government of Guyana, 1992).

d) The Amerindian Research Unit at the University of Guyana has been conducting studies on the relationship of native Amerindians to their environment.

e) At present, all B.Sc. final year students are required to research topical environmental issues.

f) The 'Flora of the Guianas Project' is undertaken by the Biology Department of the Faculty of Natural Sciences in collaboration with the Smithsonian Institute. This project involves identification, documentation and monitoring of the floral diversity of some of the forest areas. One of the results of this project is the 'Centre for the Study of Biological Diversity', which was established in 1991. This Centre documents the flora and fauna of forests, keeps inventories of plants, animals and insects, and relates this data to their ecosystems. It is envisaged that part of the work of the Centre will be in the area of environmental education both at the formal and non-formal levels.

g) The Bachelor of Education programme at the University of Guyana offers some degree of environmental education training to primary and secondary school teachers. In particular, environmental education concepts are introduced in the curricula for Social Studies and Home Economics in courses such as 'Understanding the Community', 'Interpreting the Community', 'Introduction to Social Studies', 'Issues in Social Studies' and 'Nutrition in the Community'.

Green curriculum content

Generally, four broad categories are used to group environmentally related disciplines at university level. These categories are: Environmental Design; Environmental Health Sciences, Environmental Conservation and Management, and Human and Social Ecology. It must be mentioned that there are overlaps in these categories. Nevertheless, using the subject headings, environmental education at the University of Guyana (UG) would appear as in Figure 2. It should be mentioned that the dominant approach to environmental education employed by the University of Guyana, is the incorporation of environmental themes - an environmental dimension - into some of the existing disciplines offered by the institution.

Figure 2. *'Environmentally related disciplines' offered by the University of Guyana*

CATEGORY	DISCIPLINES
Environmental Design	Architecture, Civil Engineering and Geography
Environmental Health Sciences	Occupational Health and Safety Public Health Environmental Health
Environmental Conservation and Management	Agriculture, Biology, Chemistry, Environmental Studies, Forestry, Geography, Mechanical Engineering
Human and Social Ecology	Education, Law

Appendix 2 shows the specific 'environmentally oriented courses' offered by the University of Guyana. A number of conclusions may be drawn. According to Appendix 2 'environmentally oriented courses' are dominant in the disciplines of Agriculture, Biology, Civil Engineering and Geography, and less dominant in disciplines such as Architecture, Chemistry, Education, and Health Sciences. There is need to point out, however, that the incorporation of environmental education is not entirely new to the disciplines of Agriculture, Biology, Geography, and Civil Engineering. What seems to be lacking, however, is a more systematic approach to the integration of environmental education into all the disciplines. Moreover, 'Greening of the Curriculum' is notably absent in the Arts, Economics, History, Physics, Sociology and Management Studies.

As pointed out earlier, the study of the environment necessarily involves all disciplines, as Smith (1993) notes that "for the scientist it may be represented as a system in which the input is provided by earth life and social sciences; mathematics and systems analysis have a central processing role; and the output leads to planning, designing and engineering". The writer adds that "the same pathway must be followed in parallel by exponents of the arts, whose capacities for observation, for perception of pattern, and for creative expression ... are as necessary to the educated citizen". It follows therefore that no single traditional discipline can provide students with a full understanding of the environment in its totality.

In the context of Guyana, the incorporation of environmental education into the Social Sciences, particularly, Economics, Management, Sociology and Politics, should be given serious attention - the reason being that most of the country's policy-makers are Social Science graduates. Guyanese economists, for example, should promote greater appreciation of the way the productive process is organised, as well as the distribution and consumption of goods and services meeting society's needs. Perhaps such appreciation would lead to the assessment of the environmental impacts of choices made (Connect, 1992).

Some examples of 'environmentally oriented course' content

Geography: Introduction to Human Geography
This first year 'service course' is offered to most undergraduate students from various faculties of the University. It is approached through a series of man-environment studies, examining such aspects as the dynamics of human population, agricultural and urban systems, cultural diffusion, resources and conservation, territories and spatial inequalities.

Civil Engineering: Water Supply / Public Health Engineering
This course is offered to final year students. It examines the issues of water and waste, physical, chemical and biological treatment processes, processing of sludge, disinfection and re-use of water.

Agriculture: Crop production
Crop production is a final year course that looks at crops in terms of their origin and distribution, soil and climate, seed variety and sowing, manuring, irrigation and drainage, weed control, crop management in terms of understanding crops and their environmental complexities, crop technology, and weed control.

TEACHING METHODS

Generally, the teaching methods employed by the University of Guyana are lectures, tutorials, experiments, and to a limited extent, field studies, although the former is more dominant. Even then, the delivery of 'good lectures' are hampered by the lack of recent publications and the unavailability of equipment. Due to the constraints of time, adequate budgetary allocation and poor facilities, it is very difficult to employ other methodologies such as field studies and research projects. In fact, the involvement of both students and teachers in research projects is something to be desired, although there are few exceptions.

CHALLENGES AND SUGGESTIONS

- Communication and co-operation among university departments / disciplines: environmental education should not be considered as a 'discrete subject'. It should be promoted as general education involving all disciplines.

- Broadening of pedagogical techniques: more emphasis should be placed on *learning* than *teaching* if the ultimate goal of environmental education is to make environmentally responsible citizens. The over-reliance on the lecture method at the University of Guyana, should gradually change to accommodate the use of field studies, discussions, projects and research.

- Integrating environmental education in the Arts and Social Sciences: to achieve this objective, the University of Guyana should first establish a University policy on environmental education with very clear objectives, followed by the establishment of a co-ordinating body comprising representatives from all disciplines and departments. This will provide the forum that will generate discussions, so that a general agreement on the best way to undertake environmental education can be found, as well as fostering interdisciplinary collaboration. The task of this group would also be to prepare a plan / programme for the implementation of the policy and to provide advice to individual course tutors, as well as monitoring the programme and conducting regular evaluations.

- Academic acceptability: all members of the academic staff of the University of Guyana should see the importance as well as the relevance of environmental issues to their specific discipline.

- Training: it may be in the best interests of the University of Guyana to provide short training programmes on how to incorporate environmental education into universities' curricula, for all academic staff. This will help to build the self confidence of the tutors.

- Networking: the University of Guyana should network with other Caribbean universities, particularly the University of the West Indies, to promote exchange of innovative ideas, information and resource materials as well as collaborative research.

- Creating an environmentally responsible citizenry: the University, through its public awareness programmes, should help to create a society in which people will be so environmentally aware of the needs to conserve and protect the environment that they will accept and fully support environmental legislation, monitoring and restrictions, viewing them as part of the natural order, designed for their benefit.

- Environmental Education and Research Unit: the University of Guyana should set up an Environmental Education and Research Unit (as part of the Environmental Studies Unit) to conduct research and provide technical assistance in the field of environmental education.

- Convening meetings: regular meetings for the academic staff employed by the University of Guyana should be held to facilitate discussions on common problems inherent in delivering environmental education across the University curricula. These meetings should also promote the exchange of experiences.

It should be noted that the above list of challenges has not been placed in any order of priority, but that they all need to be met in order to fulfil the role of the University in the national effort to achieve sustainable development in Guyana.

CONCLUSION

Overall, it can be concluded that, similar to other developing countries, Guyana has been influenced by developments since the Belgrade Conference, and the University of Guyana is currently moving slowly towards 'greening' the entire curriculum. Although it may not be possible to meet all the challenges stated above, the University of Guyana can begin to review the existing situation, in terms of the possibilities and the problems (both administrative and pedagogical). This will facilitate the gradual change in the status of environmental education at the institution. Moreover, in order to effectively perform its role in society as outlined by the writer, the University of Guyana will have to make 'greening the curriculum' a priority for the institution, as well as a challenge for the 21st century. After all, the University of Guyana, as any other university, should be a service institution for society.

REFERENCES

Ali Khan, S. (1991) *Greening the Curriculum, Working Document.* Committee of Directors of Polytechnics and World Wide Fund for Nature.

Ali Khan, S. (1993) *The Environmental Agenda -Taking Responsibility. Promoting Sustainable Development through Higher Education Curricula.* University of Hertfordshire, Hatfield.

AUDES (Association of University Departments of Environmental Sciences in Europe) (1995) *Newsletter, Issue 1/March 1995.* Utrecht.

Boom, E. K. (1994) 'Introduction to Seminar' In *Environment and Development Education for Developing Countries, Proceedings of a Seminar held in Accra, Ghana, August 2 - 4.* Department of Human Ecology, Brussels.

Bynoe, P. and Leal Filho, W.D.S. (1995) 'Current Trends in Education and Public Awareness in Guyana' In *International Journal of Environmental Education and Information.* (In Press).

Connect (1992) *Changing Minds: Earthwise, Connect. A selection of articles, 1976-1991.* UNESCO, Paris.

Emmelin, L. (1975) *Environmental Education at University Level.* Council of Europe, Strasbourg.

Government of Guyana (1980) *Constitution of Guyana.* Georgetown.

Government of Guyana (1992) *Development Trends and Environmental Impacts: A Country Report submitted to the United Nations Conference on Environment and Development.* Georgetown.

Government of Guyana (1994a) *National Environmental Action Plan.* Georgetown.

Government of Guyana (1994b) *Environmental Protection Bill (Discussion Draft).* Georgetown.

Howell, C. (1994) *Education for Sustainable Development- Environmental Education Programmes in the Caribbean.* Caribbean Conservation Association, Barbados.

Francis, G. (1985) 'Perspectives in Graduate Level Environmental Management Courses'. In *Environments* 15 (3), pp 29 - 41.

Irvine S and Manns H (1994) 'Towards a greener curriculum', *Series 9, Experiential Learning.* University of Northumbria.

Jarvis, C. (1981) 'Opening Remarks'. In UNESCO *National Experimental Training Workshop in Environmental Education Report.* Ministry of Education, Georgetown.

Leal Filho, W.D.S. (1990) 'A Comparative Analysis of Environmental Education in the United Kingdom and in Brazil' *In Higher Education in Europe.* 15(4), pp 45-54.

Smith, D. (1993) 'Implementing the Role of Higher Education in Environmental Education' *In Learning for Life: Environmental Education in Scotland Proceedings of a Symposium.* organised by the Royal Society of Edinburgh, Edinburgh.

Sterling, S. (1992) 'Mapping Environmental Education- Progress, Principles and Potential' In Leal Filho, W.D.S. and Palmer, J.A.(Eds.) *Key Issues in Environmental Education 1.* The Horton Print Group, UK.

The Institution of Environmental Sciences (1993) *A Framework for Environmental Education Across the Further and Higher Education Sector.* An IES Position Paper.

UNESCO (1977) *Trends in Environmental Education.* UNESCO, Paris.

UNCED (United Nations Conference on Environment and Development) (1992) *The United Nations Conference on Environment and Development: A Guide to Agenda 21.* UN Office, Geneva.

Universityof Guyana, *University of Guyana Bulletin,* 1988-1989, University of Guyana, Georgetown.

Ventura, F. (1994) 'Environmental Education- The Malta Experience' In Leal Filho, W.D.S. (Ed.) *Environmental Education in Small Island Developing States.* The Commonwealth of Learning, Vancouver.

World Bank (1992) *Draft Economic and Financial Policy Framework for Guyana, 1992 -1994.* World Bank, Washington.

APPENDIX 1

B. Sc. Curriculum for Environmental Studies: programme outline

First Year

BIO 111	-	Introductory Biology 1
BIO 121	-	Introductory Biology 11
ENG 115	-	Introduction to the Use of English
ENG 125	-	The Use of English
ENV 111	-	Environment and Development
ENV 121	-	Basic Ecological Concepts in Relation with Environment
GEO 111	-	Introduction to Human Geography
GEO 113	-	Maps and Diagrams
MTH 111	-	Algebra
MTH 122	-	Calculus

SECOND YEAR

AST 111	-	Introduction to the Indigenous People
CHM 111	-	General Chemistry 1
CHM 121	-	General Chemistry 11
CST 111	-	Introduction to Microcomputers
CST 121	-	Software Application on the Microcomputer
ENV 221	-	Environmental Issues in Guyana and the Caribbean Region; Global Environmental Issues
GEO 213	-	Field methods and Techniques in Environmental Science and Geography
GEO 112	-	Introduction to Physical Geography
MST 111	-	Basic Statistician

THIRD YEAR

AGR 221	-	Agricultural Microbiology
AGR 353	-	Fundamentals of Soil Science
ARH 303	-	Environmental Studies 1
BIO 211	-	Survey of the Invertebrates
BIO 210	-	Survey of the Cryptogams

69

	-	Survey of the Vertebrates
BIO 220	-	Survey of the Phanerogams
BIO 403	-	Ecology
BIO 455	-	Plant Ecology
BIO 402	-	Applied Biology
CHM 211	-	Organic Chemistry
CHM 221	-	Inorganic Chemistry
CIV 212	-	Geotechnical Engineering
CIV 433	-	Public Health Engineering 11
CIV 323	-	Public Health Engineering 1
	-	Issues in Education and Development 1
EFN	-	Issues in Education and Development 11
GEO 203	-	Tropical Environmental Systems
GEO 305	-	Biogeography
GEO 404	-	Geomorphology
	-	Physics for Environmental Studies
WST 200	-	Women and Development
	-	Water, Health and Environment
	-	Environmental Toxicology
	-	Atmospheric Science
	-	Aquatic Science
	-	Nature Conservation and Management
	-	Practical Environmental Chemistry
	-	Introduction to Geographic Information Systems (GIS)

FOURTH YEAR

ENV 411	-	Environmental Impact Assessment
ENV 412	-	Public Policy
ENV 413	-	Environmental Economics
ENV 414	-	Environmental Law
ENV 415	-	Environment and Education
ENV 416	-	Research Project

APPENDIX 2 *The existing "environmental oriented courses" offered by the University of Guyana*

DISCIPLINE	ENVIRONMENTALLY ORIENTED COURSE
Architecture	Environmental Studies 1
	Environmental Studies 11
Agriculture	Principles of Livestock Production
	Principles of Crop Production
	Agricultural Entomology
	Agricultural Engineering
	Applied Soil Science 11
	Crop Production
Biology	Introductory Biology
	Survey of the Animal World
	Survey of the Plant World
	Ecology
	Plant Ecology
Chemistry	Environmental Chemistry
	Organic and Agricultural Chemistry
Civil Engineering	Geotechnical Engineering
	Hydrology/Hydraulic Engineering
	Water Supply/Public Health Engineering
	Water Resources/Drainage and Irrigation
Education	Understanding the Community
	Interpreting the Community
	Introduction to Social Studies
	Issues in Social Studies
	Nutrition in the Community
Environmental Health	Water, Sewage and Solid Waste Management
Forestry	Introduction to Forestry
	Forest Protection
	Silviculture and Dendrology
Geography	Introduction to Physical Geography
	Introduction to Human Geography
	Tropical Environmental Systems
	Biogeography
	Urban Planning
	Geomorphology
	Aspects of Rural Development
	Regional Planning and Development
Health Science	Environmental Studies
	Occupational and Institutional Health
	Water, Sewage and Solid Waste Management
Law	Environmental Law
Mechanical Engineering	Soil and Water Conservation
	Alternative Economic Development

Chapter 5

ENVIRONMENTAL EDUCATION IN THE FORMAL SYSTEM: THE TRAINING OF TEACHERS

Joyce Glasgow
Richmond Vale
Runaway Bay
Saint Ann
Jamaica

INTRODUCTION

The essential needs of the formal education system for fulfilling its expected role in environmental education are basically the same as those required for the proper delivery of any educational programme, namely:

* political and economic support at the state level;
* a programme or curriculum through which the desired goals may be achieved
* teachers and other human resources;
* materials and equipment;
* local community support.

Of these 'essentials' the right teachers are absolutely crucial to the environmental education enterprise, especially as they are expected to be models of the values they try to encourage in their students.

PREPARING TEACHERS FOR ENVIRONMENTAL EDUCATION

If teachers are so essential to the success of environmental education, as has been espoused since Belgrade, then their thorough preparation for the task should receive high priority. In planning for such preparation, one has to take into account both the qualities which should characterise environmental education, as well as the general aims of environmental education. In sum, these aims may be expressed by saying that the desired end products of the educational process are 'world citizens'. Such citizens feel at one with their environment, are self-disciplined, and see themselves as guardians of the local, national and international good, willing to go beyond questioning to plan and implement action to improve the health of the planet.

The expected qualities of environmental education may be deduced from the now accepted general principles for the guidance of environmental education which emanated from Tbilisi in 1977. The directives given, in terms of traditional educational practice, may be regarded as revolutionary. They direct that environmental education should:

a) be holistic in nature and lifelong in duration;
b) go across boundaries of time, place, nations and peoples, and intellectual disciplines;
c) be concerned with real-life issues, encouraging responsible citizenship and plans for sustainable development and growth;
d) promote participatory learning using a diversity of approaches, and provide practical first-hand experiences;
e) develop in the learner diagnostic, problem solving and evaluative skills.

Except for the skill development in (e), these directives are in stark contrast to the common trends in traditional education, which tends to be fragmentary, rather than holistic in outlook, to be constrained by boundaries of one sort or another and to separate schooling from real life.

With these two aspects in mind, one could enunciate the following objectives for the training of teachers for any level of the formal system. Such training should:

• give teachers a holistic view of nature, man's place in it, and his responsibilities to it for his own and future generations;
• help teachers to understand the complex linkages and interdependence between environmental integrity and sustainable development;
• equip teachers with sufficient knowledge about the total environment, the problems engendered by human socio-cultural development and their possible solutions, to enable them to handle environmental education competently;
• help teachers to recognise the interdisciplinary nature of environmental education;
• help teachers to conceptualise and integrate this interdisciplinary dimension into their teaching;
• equip teachers to use a problem-solving, activity-oriented, participatory approach suitable for their students;

- enable teachers to identify, wherever they work, the resources (including human beings), for focusing the attention of their students on environment and development situations and problems;
- enable teachers to use these resources to devise ways of interweaving the environmental dimension into the curriculum;
- help teachers acquire those values of responsible citizenship and care for the environment which will motivate them to take appropriate action where necessary to protect and enhance the environment;
- help teachers to realise the necessity for continuous updating of their knowledge and skills;
- give teachers the confidence to apply new methodologies, and to use new content on a continuing basis.

(modified from Glasgow, 1994)

An examination of these objectives will show that teacher training for environmental education must seek to provide teachers with a double set of competencies. There are the general foundation competencies which environmental education shares with all other aspects of education, as well as competencies of knowledge, skills, attitudes and values which are unique to environmental education (see Wilke *et al.*, 1987). Figure 1 summarises these competencies.

It would be difficult for any one teacher education programme to prepare teachers adequately in all the competencies, and in any case, the nature of environmental education requires lifelong preparation. Perhaps, however, the important target should be that the programmes seek to address all the competencies mentioned. The stress given to each facet will depend, *inter alia*, on the academic and professional background of the particular audience, the school curriculum, and the level for which the teachers are being trained, or other rationale for offering the course.

A MODEL FOR TEACHER TRAINING IN ENVIRONMENTAL EDUCATION

Figure 2 suggests an integrated plan for teacher training in environmental education suitable for the Caribbean. The model would ensure that every teacher in training in - and for - the formal system of the region, would receive a modicum of exposure to environmental education. The exceptions would be those employed at the university level where only employment requirements could have any kind of similar effect. Supporting the provisions for teachers actually attending the teacher education institutions, are arrangements for classroom teachers to receive environmental education instruction through vacation courses, regular day-release seminars and workshops, lecture series, peer teaching and short term (not more than three months) university level programmes. Details of the plan are given in Appendix 1.

The plan is workable because the formal systems throughout the region are similar, and because the overall geographical area covered is small. The principles may, however, be adaptable to other discrete educational administrative blocks.

It is assumed that the content of the courses titled ENVED I to VI will satisfy, to varying degrees and levels, the competency requirements outlined in Figure 1. That is, they will address:

A. The philosophy and aims of environmental education

B. Ecological foundations:
- Living things in their habitat - species and ecosystems
- The impact of human activity on ecosystems -
    human beings and the land
    human beings and water
    human beings and the atmosphere
    human beings and energy and minerals
    human beings and their communities
    science, technology and development
    the global commons and international issues

C. Teaching methodologies and techniques with an emphasis on those which will best promote the participatory, decision-making, values clarification and personal nature of environmental education. For example:
    :simulation, gaming, role-play
    :situation analyses
    :use of the performing arts
    :use of projects and field trips
    :research from primary and secondary sources.

D. Methods of evaluation in environmental education

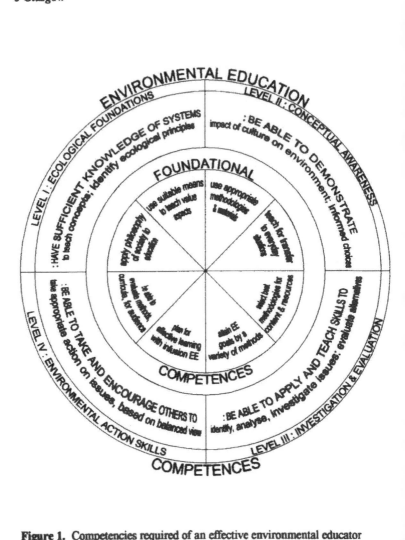

Figure 1. Competencies required of an effective environmental educator (modified from Wilke *et al.*, 1987)

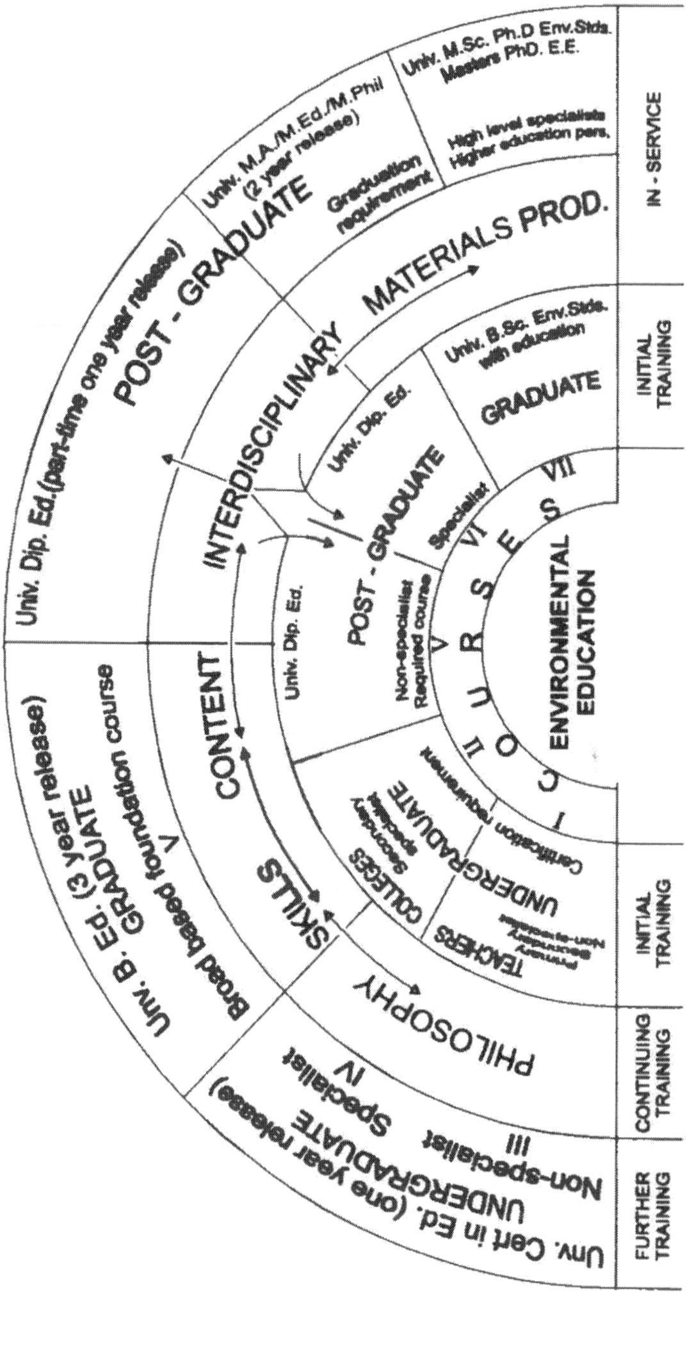

Figure 2. Suggested scheme for the training of teachers in environmental education for the formal system

THE REALITIES

A possible model for the ideal in teacher education provisions in the Caribbean has been presented. It is necessary now to look at the situation as it exists for comparison, and to extend this comparison somewhat by a brief look at a few locales outside of the Caribbean.

The Caribbean

There are several opportunities for teachers or prospective teachers to upgrade their technical environmental knowledge and skills. The universities in the region offer a variety of courses and degree programmes at Bachelor and post-graduate level, which deal with several aspects of environmental science, resource management and the connected issues, and the opportunities for research are numerous.

Structured teacher education programmes are, however, another matter and research into environmental education practices which should inform the enterprise is lacking. The University of the Virgin Islands, through its Eastern Caribbean Centre, does conduct teacher training exercises, and at the Centre for Resource Management and Environmental Studies (CERMES) at the Cave Hill campus of the University of the West Indies, there was a 'one shot' experiment some four years ago wherein a selected group of six teachers combined the technical work for the post-graduate diploma in environmental management with a course in environmental education. This potentially useful model was, unfortunately, discontinued. Of import for the future, however, is the fact that, beginning in 1995, at its Mona campus, the University of the West Indies will offer a course in environmental education as part of its Masters programme in teacher education.

In the teachers' colleges, where the initial training of teachers for the primary, and in some cases, secondary system, takes place, environmental education is left mostly to the devices of tutors of the sciences and social studies, although prescribed curricular content allows for environmental education. It is also true to say that in terms of taking steps to improve their ability to encourage attitudinal changes expected in true environmental education, the universities have not looked at themselves. This is in spite of their technical expertise, and the fact that the Consortium of Caribbean Universities, which operates under the umbrella of the UNEP Caribbean Environment Programme, does include in its programme the training of faculty in interdisciplinary teaching methodologies.

78

At all levels, however, there have been curriculum innovations which seek to make the environment an important issue, but provisions for teacher education are never prior to, nor concurrent with the fact, but rather after the fact.

The Maltese experience

Of Malta, Ventura (1994) states that systematic formal environmental education is still in its infancy, and that there is still room for experimentation. A place for environmental education in the system is, however, assured, as since 1988, there have been statutory provisions for it. Curriculum innovations in environmental education have been introduced at all levels of the system, from primary through to post-secondary. To 'match' this, environmental education in teacher education has been available since October 1993 for teachers at the primary level in the B.Ed. (Hons.) programme at the University of Malta. Arrangements were yet to be made to prepare teachers for handling a compulsory course in environmental studies for the last three years of secondary school, slated to begin in October 1993.

Ventura further emphasises that there is a commitment to environmental education at the tertiary level both in teaching and research at the undergraduate and post-graduate levels. He sums up by saying that since 1987

"...various faculties and institutes of the University of Malta, the Foundation for International Studies, and the Malta Council for   Science and Technology have organised courses, workshops and conferences, and set up research projects regarding the terrestrial and marine environment, environmental economics and environmental law."

The statement suggests that the considerable activity at this level stresses a technical knowledge of environmental concerns, but one is left to wonder about the 'education' component of environmental education as it affects how university staff teach.

Therefore, as in the Caribbean, one is left with the twin impressions of a time lag between curriculum innovations at school level and teacher preparation to cope with them, and an uncertainty as to self-examination in environmental education by university staff who are at the top of the chain.

Nigeria

Agbola (1993) in reviewing the situation in Nigeria, points out that even in the absence of a coherent environmental education policy or strategies to realise such a policy, children in primary and secondary schools are exposed to an appreciable measure of environmental knowledge and discussion of issues. At the university level, however, specialisation is, as expected, the norm. Whereas students in the faculties of technology, science, agriculture and forestry, and social sciences would have been exposed to discussions on environmental issues, this would not necessarily be true of most students in the arts, education and other courses.

Of the preparation of teachers for the primary system in the teachers' colleges, Agbola opines that not only are a "learner's interest in exploring the environment and in coping with the changing environment' stimulated, but students are trained in the 'principles and practice' of communicating with the children they will teach" (Agbola, 1993).

At the secondary level, however, provisions are not so pervasive. In the colleges of education, student teachers' exposure to environmental education would vary depending on their subject combinations. Students in the sciences or social sciences are likely to have a significant infusion of environmental education, while their counterparts in arts and business education would not. Agbola points to the undesirability of this situation, since teachers sometimes have to assume responsibilities outside of their areas of specialisation.

He is, however, concerned that the efforts at primary and secondary level have not been effective in making a difference to the attitudes of the students who have passed through, nor enough to make an impact on practice. The question is raised as to the teaching methodologies used; the suggestion is that these have only involved students in what Agbola terms 'cognitive based' environmental education. If so, then one basic tenet of environmental education would not have been upheld. Environmental education must not only be participatory - it must involve a diversity of approaches, and provide practical first-hand experiences which are topical and relevant to the real life situations of the learners.

Kenya

In Kenya, Wanaswa (1993) states that although there is no comprehensive policy on environmental education, on the directive of a Ministry Paper environmental studies have been taught at all school levels, and formed a part of teacher training curricula. As in the Nigerian situation discussed above, environmental education is infused into curricula at the primary and secondary school levels, and in the training of teachers for the primary schools. Teachers being prepared for the secondary system in the Diploma Colleges, if their options involve the natural or social sciences, will benefit from a treatment of environmental themes relative to their area, both in content and methodology. Additionally, however, all students in these colleges are required to do a special course in environmental education. This speaks well for the potential for environmental awareness in graduates of these colleges.

Wanaswa points, however, to the fact that teachers have not established the relationships with the communities in which they work, which would allow them to avail themselves of the opportunities for facilitating actual environmental experiences, and of using local resources, human and otherwise. This, however, is an important facet of environmental education that should be stressed in their training, by example. She further points to the necessity for in-service training for the large numbers of teachers who would not have been exposed to environmental approaches or have been sensitive to the issues raised in the curricula they are called upon to follow in the schools.

England and Wales

The situation in England and Wales with regard to environmental education seems to be 'in flux' and variable in a system which is more decentralised with respect to administration than any of the four cases so far discussed. Hale (1993) points out that -

"The recognition of environmental education as a cross-curricular theme in the National Curriculum was of great significance in terms of putting environmental education on the 'curriculum map'."

Unfortunately, however, the "sense of importance which teachers attach to environmental awareness and skills" is diminished because environmental education is not examined.

Hale also opines that it is apparently assumed that teachers have the skills, experience and time to deal effectively with introducing cross-curricular themes. On the contrary, however, it should be realised that to treat environmental education as a cross-curricular theme demands a firm grounding in environmental education of all teachers participating in the exercise. This facilitates planning for an appropriate sequence of learning experiences.

The development of initial training programmes to support environmental education in the National Curriculum is, however, not uniform, although the Council for Accreditation of Teacher Education (CATE) has the task of ensuring that some environmental education is part of teacher training. With regard to in-service training, INSET directives ignore environmental education. Additionally, education advisers now have little time to structure in-service courses because of the changing structure of the local education authorities. There has also been a reduction in the support, in particular that from local authority environmental and field study centres.

Thus, once again, there is the picture of curriculum innovation without proper provision for teacher preparation to cope with the new trends, standards and responsibilities.

THE COMMON THREADS

The common threads which seem to emerge from this brief comparison are the following.

1. Primary and secondary school curricula appear to have adequate coverage of basic environmental knowledge and issues. Children do not, however, seem to have internalised the attitudes and values of environmental education which should have been stressed in the teaching, as they have not put these into practice as school leavers.

2. There is, generally, a time lag between curriculum innovation at the school level and teacher preparation to deal with this.

3. Even where teacher training provisions are in place, the indications are that the delivery of environmental education in schools has not had the effect desired (see 1). Two very feasible propositions have been made to try to explain the shortcoming. Both concern teaching methodologies. The first suggests that teachers may not have provided the necessary practical experiences. The second intimates that the experiences provided may not have been personal and relevant enough to the learners. Both characteristics are recognised keynote attributes of environmental education.

4. More attention is paid to providing initial teacher training in environmental education than to accommodating in-service training.

5. Scant heed is paid to environmental education in technical and vocational education.

6. University staff need to recognise that they should be trained for the education competencies, even though they may be the acknowledged technical experts in areas of environmental concern or knowledge.

BRIDGING THE GAPS - THE FUTURE

The time lag

"Formal systems are slow to adapt, and tend to package their knowledge in course-size parcels for sale to limited numbers of students" (Crocombe, 1994).

It is incumbent on curriculum developers and planners to take steps to close the time gap between curriculum innovations and the teacher preparation to support them. Funding, time available (and often these two are linked), crises situations - any number of constraints may stand in the way of teacher preparation before the fact. Why, however, cannot the plans for teacher training be concurrent with the designing of the curriculum? If teachers and teacher educators are involved at the designing stage, these participants would be the first line of trainees, as it were, and subsequently the first line of trainers. In this way, the initial steps in teacher preparation would be a step ahead of implementation in the institutions, instead of lagging far behind.

The difference to implementation that the absence of this time lag can make is illustrated by the case of Madagascar. One of the factors that no doubt accounts for the much publicised success of the World Wildlife Fund's education programme in Madagascar, is the fact that the structures necessary to 'get off the ground' were put in place relatively early. Several activities were going on at the same time - introducing environmental education in the formal school curriculum; training teachers to teach the 'new' ideas producing and using teaching materials familiar to the children, for example, Malagasy-language school manuals, traditional songs, stories and poems, and building environmental education centres. So curriculum, teachers' and resource materials were to hand, and community support and involvement was good.

One recognises that although the project has been developed, and is run by the Malagasy people, without international funding and expertise, all the desirable components of the programme could not have been put in place so expeditiously. Nonetheless, the principles are instructive.

Community relations

The possible absence of practical activities, and of experiences personal to the learner in the environmental education teaching / learning situation can largely be traced to a lack, on the part of the teacher, to recognise that the community is the source of both the foci for environmental education and of the resources to accomplish its ends. Such foci are, for example, community ecology relationships, community customs and cultural practices, community problems and the search for solutions. The resources are the physical and biological elements, the human and cultural attributes, including traditional knowledge and skills in environmental resource management and health. Community-based environmental education would guard against what Agbola (1993) aptly terms 'cognitive-based' environmental education.

Teacher educators should emphasise this aspect of environmental education by example in the training programme. Elliott (1991) suggests that the practical investigation of real-life problems not only helps to upgrade teachers' knowledge base, but allows them to "gain practical competence in complex human situations requiring wisdom as a quality of judgement and decision-making with respect to their classroom practice."

Continuous training

Initial training for new teachers should be followed by in-service training on a regular basis. As Chambers (1991) suggests, in the time available for initial training,

"...only environmental specialists could begin to approach the knowledge, skills and values base required of a teacher of environmental education immediately after initial training."

In-service training needs also to be designed for the bulk of classroom teachers who will not have had the benefit of initial training in environmental education during their 'college years.'

Environmental education in technical and vocational education

It is paradoxical that although the daily activities of the individuals being catered for in technical and vocational education are likely to have a considerable impact on the environment, little is done to ensure their awareness of this fact, except perhaps in colleges of agriculture. Quite apart from this consideration of the 'outer environment', environmental education in technical and vocational education should impart an awareness of the organisational, equipment or process related, and product-related risks associated with any particular area of activity. It goes without saying that teachers should be very highly trained, so as to be able to guide students without mishap.

The role of the Universities

Universities as a group have attempted to encourage the attitudinal path of environmental education, but the audience of this attention has largely been primary and secondary level students and their teachers. Little attention has been turned on themselves. This lack needs remediation. The universities in any region, represent the largest intellectual repositories, and have the best facilities that can be afforded. They must lend leadership in all areas.

The body of technical knowledge 'about' the environment is staggering, but, in the Caribbean at any rate, universities have not explored the socio-cultural environment to the same extent. Neither have they given enough attention to research in environmental education practices. Research is needed to inform future directions in teacher education.

Finally it is necessary to make three observations. Firstly, in any context, but especially where financial and personnel considerations are paramount, it is imperative to maximise the use of whatever resources are available for teacher education in environmental education. Strategies include peer teaching, national and regional collaboration and sharing, and the use of existing distance education networks.

Secondly, minimum acceptable standards for teacher training in environmental education should be set and required.

Thirdly, despite the fact that innovations take a very long time to 'make a difference', the formal education system, catering as it does to pre-school through to young adult audiences, needs to remain the focal point for environmental education. Gains take time to effect, but are likely to be more permanent. For success in environmental education, however, the system must avail itself of the expertise and other resources in the community and non-formal system. This is especially true of the teacher education undertaking.

ACKNOWLEDGEMENTS

Information on teacher training in Nigeria, Kenya and England and Wales was drawn from three chapters in Leal Filho, W. (Ed) (1993) 'Environmental Education in the Commonwealth'. The Commonwealth of Learning. The relevant chapters are 'Environmental Education in Nigerian Schools' by Dr. Tunde Agbola, University of Ibadan, Nigeria; 'The State of Environmental Education in Kenya Past, Present, and Future', by E.A. Wanaswa, Kenya Institute of Education, Kenya and 'Environmental Education in the National Curriculum of England and Wales' by Monica Hale, London Guildhall University, United Kingdom. Similarly, the chapter entitled 'Environmental Education - The Malta Experience' by Frank Ventura, in Leal Filho, W. (Ed) (1994) 'Environmental Education in Small Island Developing States'. Commonwealth of Learning supplied the information on Malta.

REFERENCES

Agbola, T. (1993) "Environmental Education in Nigerian Schools" In Leal Filho, W. (Ed) *Environmental Education in the Commonwealth.* Commonwealth of Learning, Vancouver.

Chambers, W. (1991) "Environmental Education and Initial Teacher Training" In *Conference Report,* International Conference on Environmental Education, Manchester UK. The Department of Education and Science.

Crocombe, R. (1994) "Environmental Education in the South Pacific" In Leal Filho, W. (Ed) *Environmental Education in Small Island Developing States,* Commonwealth of Learning, Vancouver.

Elliott, J. (1991) "Teacher Training for Environmental Education Developing Community Focused Environmental Education Through Action Research in Schools" In *Conference Report,* International Conference on Environmental Education, Manchester, UK. The Department of Education and Science.

Glasgow, J. (1994) *Environmental Education Curriculum Guide for Pre-Service Teacher Education in the Caribbean - Environmental Education Series 39.* UNESCO, Paris.

Hale, M. (1993) "Environmental Education in the National Curriculum of England and Wales" In Leal Filho, W. (Ed) *Environmental Education in the Commonwealth.* Commonwealth of Learning, Vancouver.

UNESCO, (1978) *Final Report. Intergovernmental Conference on Environmental Education. Tbilisi 1977.* UNESCO,. Paris.

UNESCO, (1980) "ENVED (Caribbean 4) Strategies for the Training of Teachers - Environmental Education for Primary and Secondary Schools and Teacher Education Institutions". *Discussion Guide for the Sub-regional Training Workshop on Environmental Education for the Caribbean, Antigua, 1980.*

UNESCO, (1983) *Trends in Environmental Education since the Tbilisi Conference.* UNESCO, Paris.

Ventura, F., (1994) "Environmental Education - The Malta Experience" In Filho, W., (Ed) *Environmental Education in Small Island Developing States.* Commonwealth of Learning, Vancouver.

Wanaswa, E.A. (1993) "The State of Environmental Education in Kenya Past, Present, and Future" In Filho, W., (Ed) *Environmental Education in the Commonwealth.* Commonwealth of Learning, Vancouver.

Wilke, R., Peyton R., Hungerford, H., (1987) *Strategies for the Training of Teachers in Environmental Education: Environmental education Series 25.* UNESCO, Paris.

World Wildlife Fund, *Mission for the 1990s.* WWF International, Switzerland.

APPENDIX 1

Details of teacher training scheme

Initial and further training

Note, The terms (1) specialist and (2) non-specialist are here used to indicate student teachers (1) with subject options in the natural or social sciences and (2) with other options. Initial training programmes are for minimum certification. Further education programmes allow upward mobility in the system. Continuing education allows teachers presently in the classroom to cope with new developments.

COURSE OFFERINGS

Course: ENVED I
Course Level: Undergraduate
Teacher Level: Primary Secondary non-specialist (Initial)
Institution: Teachers' College
Aim: Introductory course to enrich reputedly poor environmental science background of student teachers, and enable non-specialists to add environmental direction in presenting their subjects
Implementation: Require for certification. Teach by inter-disciplinary team.

Course: ENVED II
Course Level: Undergraduate
Teacher Level: Secondary specialist (Initial)
Institution: Teachers' College
Aim: More in-depth course than ENVED I for teachers who will bear the brunt of the thrust in the secondary schools. Course will enable student teachers to present a thematic, flexible and integrated approach to environmental education in their curricula.
Implementation: As for ENVED I.

Course: ENVED III
Course Level: Undergraduate
Teacher Level: Certificate in Education (Further) Non-specialist.
Institution: University

Aim: Mini-background course with socio-cultural emphasis to heighten awareness in these experienced primary school teachers, and enable them to add some environmental direction to their work.

Implementation: As ENVED I and II.

Course: ENVED IV
Course Level: Undergraduate
Teacher Level: Certificate in Education (Further) Specialist.
Institution: University
Aim: A more in-depth form of ENVED II for experienced teachers to upgrade and lend new direction to their knowledge and teaching skills.
Implementation: Combine and replace some sections of current science / social studies / geography content areas. Teach through interdisciplinary team.

Course: ENVED V
Course Level: Graduate post-graduate
Teacher Level: BEd, MA / MEd. (Further) Diploma in Education (Initial) All non-specialist.
Institution: University
Aim: Broad background course for experienced college trained teachers or degree students wanting professional training (DipEd)
Implementation: Require as foundation course

Course: ENVED VI
Course Level: Post-graduate
Teacher Level: Diploma in Education specialist (initial)
Institution: University
Aim: In depth course to encompass and extend subject specialisations in the natural and social sciences, allowing teachers to broaden their scope and heighten their awareness of their environment.
Implementation: Will constitute the normal course in the subject area. Teach through interdisciplinary team.

Course: ENVED VII
Course Level: Graduate
Teacher Level: BSc (Environmental sciences) with education (initial)
Institution: University

Aim: To create a core of resource persons for all facets of the educational system, formal and non-formal.

Implementation: Cross faculty decisions between Faculties of Natural Sciences and Education to determine the selection of courses and arrangements for teaching them

Course: ENVED Higher Degree
Course Level: Post-graduate
Teacher level: MSc, MPhil, PhD All specialists Environmental education environmental studies (Further)
Institution: University
Aim: To create a core of personnel for higher education and specialists for government and private organisations.

Implementation: Use existing subject specialists supplement with visiting specialists where necessary use travel / exchange programmes.

Continuing training

Aim: To introduce to and/or reinforce in practising teachers throughout the formal system environmental education concepts and strategies for teaching them as befits teachers' level and context.

Implementation
strategies: Term time workshops, on regular day release and/or week-end system
:Block vacation courses
:Short term (3 month) university level programmes
:Regular inter- and intra- institution seminars, discussions, meetings - pooling expertise
:Radio, television programmes
:Lecture series from local and foreign specialists utilising distance education facilities at the University of the West Indies and University of the Virgin Islands.

More specifically for each level of the system, the objectives of training would be as follows.

Teachers in the Primary System

To help teachers to acquire the skills and knowledge to
1)   Infuse environmental education into the curriculum
2)   Use community resources
3)   Teach the 7-11 age group, de-emphasising the teacher.

Teachers in the Secondary System

1)   To upgrade teachers' environmental knowledge
2)   To help teachers to understand the implications of environmental education for interdisciplinarity
3)   To help teachers to acquire a knowledge of, and skills for using strategies for achieving interdisciplinarity
4)   To help teachers to recognise the importance of the community in environmental education.

Teacher Educators

The objectives for teachers in the secondary system all apply. Additionally, the training will seek:
1.   To upgrade teachers' skills in curriculum development to include environmental education
2.   To point to directions and methods for community based research.

Note: The above is a modified form of the scheme presented in ENVED (Caribbean - 4):'Strategies for the Training of Teachers - Environmental Education for Primary and Secondary Schools and Teacher Education Institutions', a Working paper prepared by this author for the UNESCO-UNEP Sub-regional Training Workshop on Environmental Education for the Caribbean, Antigua, 1980.

Chapter 6

DEVELOPING INTERNATIONAL ENVIRONMENTAL ADULT EDUCATION: THE CHALLENGE, THEORY AND PRACTICE

Darlene E. Clover

International Council for Adult Education
720 Bathurst Street, Suite 500
Toronto M5 2R4
Canada

INTRODUCTION

"Helping people to become environmentally literate requires that we stimulate knowledge, understanding, awareness, commitment, skills and then action."
(Victor Ibikunle-Johnson, 1987)

In October 1975, one of the first international workshops to focus almost exclusively on environmental education took place in Belgrade. In total, sixty four countries took part in this workshop. The result of the 10 days of deliberations was 'The Belgrade Charter', a policy statement that outlines the basic goals, objectives, concepts and philosophies, and guiding principles of environmental education (Tilbury, 1994). Daniella Tilbury refers to this event as "one of the landmarks in environmental education's history" (Tilbury, 1994). Then in 1987, the report 'Our Common Future', once again drew attention to formal and non-formal education as effective tools to help people to make personal changes in their own behaviour, and make more informed political decisions.

These two documents were important for many reasons, but, for the purposes of this chapter, their primary importance was that for the first time a need to develop environmental education methodologies, for the general public or adults, was articulated and disseminated around the world. Moreover, they highlighted the critical role education could play world-wide in helping adults better understand the political edge of the environmental crisis, something which environmental education had virtually ignored for a variety of reasons, and encouraged them to take individual and collective action to resolve environmental problems" (Orr, 1992; Tilbury, 1994).

To date, however, insufficient attention has been given to the challenge of developing environmental education for adults (Orr, 1992). David Orr (1992) argues that much still needs to be done to ensure that adults as citizens, consumers, employers, workers and parents are able to develop skills, and acquire the knowledge and awareness they need to work together to combat the environmental crisis. What is most interesting is that the vast majority of environmental education literature available such as manuals, activity guides, books, articles, and so on focus largely "upon the needs of young people and to a lesser extent, specialists in industry" (NIACE, 1993), although it is the adults of the world who are making critical decisions on a daily basis that affect the biosphere.

It is adults who control local and multi-national businesses that pollute air, land and water, financial institutions such as the World Bank that displace thousands of people to make way for hydro-electric dams, and run governments at all levels.

If education is to be an effective tool to secure change and achieve a more just and sustainable society, it will have to reach out to people at all levels in society, not just those working or studying within institutions. The major challenge facing adult educators around the world is to develop an environmental education framework based on 'The Belgrade Charter' and 'Our Common Future' that suits the special needs of adults. Adults often have less time for education, bring a variety of experiences, and have different learning methods than children to name but a few obstacles.

Environmental education practices for adults must include a socio-political analysis of the environmental crisis, be experiential, begin with adults' knowledge and experience in order to develop new knowledge, take place within specific time periods and be action-oriented in both a day-to-day sense and politically. It must elicit the support of nature as teacher and site of learning to work towards the re-development through a more sensory relationship. And it must also help to reverse human disassociation from the rest of nature, the result at least in part, of Western philosophical and scientific views based on mastery and control.

Developing and supporting the practice of non-formal environmental adult education around the globe is exciting but often quite overwhelming. However, as Wangari Maathai, of the Kenyan Green Belt Movement stated so eloquently: "You have challenged us, we shall not disappoint you."

INTERNATIONAL LANDMARKS IN ENVIRONMENTAL EDUCATION

"Nobody expects education to do it single-handedly, but the expectations generated by   major international events such as the Earth Summit at Rio and the anxieties incurred by well-publicised disasters such as Bhopal and Chernobyl...have raised the level of demand for action." (Smyth, 1995).

"The goal of environmental education is to develop a world population that is aware of, and concerned about, the environment and its associated problems and which has the knowledge, skills, attitudes, motivations and commitment to work individually and   collectively towards solutions of current problems and the prevention of new ones." (UNESCO-UNEP, 1976).

As stated elsewhere in this book, the first major international environmental gathering to draw attention to the critical role of education was the United Nations Conference on the Human Environment in Stockholm in 1972. Wheeler (1992) notes that up until that time, "in anglophone countries around the world, 'environmental education', 'environmental conservation education', and 'conservation education' existed side-by-side and were used interchangeably." But with the passing of Recommendation 96 (UNESCO-UNEP, 1976), the term 'environmental education' entered the realm "of widely accepted usage." 'Recommendation 96' states that UNESCO and member states should:

"...establish an international programme in environmental education, interdisciplinary in approach, in school and out of school, encompassing all levels of education and directed towards the general   public" (emphasis mine).

In 1975, arising from the Stockholm conference, was a key global event that took place in Belgrade and focused almost exclusively on environmental education. The result of this 10-day International Environmental Education Workshop was the adoption of 'The Belgrade Charter: A Global Framework for Environmental Education', a policy statement which outlined the gravity of the environmental situation, established a goal of action, as well as goals, objectives, audiences and guiding principles for environmental education programmes, both formal and non-formal (UNESCO-UNEP, 1976). The Belgrade Workshop emphasised that environmental education needed to be "an educational process with important ethical, economic, and political implications" (Wheeler, 1992). It also noted that one of the goals of environmental education is:

"to help individuals and social groups develop a sense of responsibility and urgency regarding environmental problems to ensure appropriate action to solve those problems."

And finally, it included by name, non-formal education and adults as part of the critical audience for environmental education:

"The non-formal education sector: including youth and adults, individually and collectively, from a segment of the population such as the family, workers, managers and decision-makers in   environmental as well as non-environmental fields."

The first ministerial-level Intergovernmental Conference on Environmental Education took place in Tbilisi, Georgia (former USSR) in October 1977. The purpose of this conference was to promote and intensify thinking, research and innovation with regard to environmental education. The Tbilisi Declaration also put forward a framework, principles and guidelines for environmental education both inside and outside the formal school system (Camozzi, 1994).

The final international conference that continued the emphasis on the need for an education that would influence personal, social, and global behaviour and permeate all aspects of life, was the United Nations Conference on Environment and Development (UNCED) held in Rio de Janeiro in June 1992. 'Agenda 21', the official government document produced contained a chapter titled 'Education, Public Awareness and Training'. This chapter outlined a framework for education that incorporates environmental and development concerns, and put a stronger focus on non-formal education. In addition, the Global Forum, the parallel non-governmental conference published the 'Environmental Education for Sustainable Societies and Global Responsibility Treaty' which will be discussed in more detail in the next section of this chapter.

Flowing from these conferences have been a variety of local, national, regional and inter-regional conferences, workshops and seminars on environmental education around the world. Tilbury (1995) notes that the past few decades have been important years for environmental education because "environmental education's holistic philosophy began to take root" and "public environmental concern continued to heighten, giving [it] a stronger impetus". Wheeler (1992) refers to this period as a "great explosion or 'quantum jump' which produced a new product, a new philosophy, a new approach: environmental education."

All the conferences and documents produced over the past few decades have drawn attention to the critical role of environmental education, the need to re-conceptualise it to suit today's world-wide crisis, and, most importantly, the political edge that underlies and as Tilbury notes, undermines all education (Tilbury, 1994; Orr, 1992). Two of the primary factors, although there are others, that undermine effective environmental education are empirical science and / or the mastery of nature, and political backlash. As noted by Tilbury (1995) and Smyth (1995), environmental education has tended to be informed primarily by positivist empirical science and its idea of besting and controlling the natural world. Science cannot cope with aspects of experience that are un-measurable such as emotions and feelings because there is "no convertible evidence of them in a test tube" (Rowe, 1990). The world's physical scientists have insisted that nature speaks only in a language of quantity, mathematics, in a mechanistic not organic way. Science supports a notion of power and control over the natural world (Rowe, 1990). Its mechanistic view of the world is seen by many ecological critiques as "one of the major factors in the deterioration of our relationship to the natural world" (O'Sullivan, 1995). A second factor is political backlash. In many parts of the world, particularly North America and parts of Europe, there has been a marked escalation in subtle and not-so-subtle attacks against environmentalism which work to pit workers against environmentalists and which have greater access to diverse forms of media that reach thousands of children and adults daily. "Environmental education has become unavoidably political" (Orr, 1992), and the need for different levels of critical adult education on the environment, increasingly more necessary.

What has become most apparent to educators around the world was that not only the goals and activities of previous environmental education had to be re-thought, but also its pedagogy (Wheeler, 1992; Deri and Cooper, 1993).

WHY ADULTS AND NON-FORMAL EDUCATION?

"More than half of all adults...are not aware of the basic environmental impact of their own activities" (NIACE, 1993).

Adult education around the world is often understood to refer solely to organised learning beyond secondary school and in terms of literacy classes, credit courses at community colleges, and extension or night classes at universities. The area of non-formal education that happens in the home, workplace, at the market and so on, is often overlooked because the learning process itself is unseen, it does not take place within classrooms, there is no 'expert' teacher and what is learned can not be calculated through testing. However, non-formal learning and education are an important component of the lives of adults and around the world, community members and adult educators are creating divergent learning opportunities for adults within their own communities.

Like much conventional education and an erroneous view of adult education, environmental education practices and activities have been geared towards young people, those studying within institutions or narrowly-focused on issues pertaining to the workplace. However, the vast majority of citizens, consumers, workers, employers, and parents are adults who make critical decisions that affect others and the biosphere everyday, and do not attend classes or experience any type of environmental education. The National Institute of Adult and Continuing Education in Britain (NIACE) outlines various reasons why investing energy and resources into environmental adult education is so important. The most crucial of these is that

"...we cannot wait for the present generation of school and college students to begin applying their newly-won environmental awareness - we must educate those who are making vital decisions now" (NIACE, 1993).

Adults not enrolled in environmental programmes in pedagogical institutions are educated primarily through their peers and/or the media. The main problem with the media is that it does not focus on process but rather events. It speaks in sound bites, snippets of information that can be over-simplistic, distorted or confused. When knowledge is wrong, Rowe (1990) writes, "wrong ideas and misdirected activities can flow from it." Increasing non-formal environmental adult education could represent an enormous potential force for change that could work as a mobilising factor to achieve environmental improvements.

Mark Burch (1994) believes that much of the traditional environmental adult education in Canada, and around the world, is very narrow and instrumental. It consists of "short, very specialised training programmes often delivered within private industry to address specific occupational issues of environmental management" and is "driven by established or impending regulatory legislation." This educational philosophy and practice developed from the erroneous assumption that environmental education (and other forms of education as well) is a one-time activity rather than a life-long process

(NIACE, 1994; Viezzer, 1986). Conventional education in the modern world, as Orr (1992) notes, "was designed to further the conquest of nature and the industrialisation of the planet. It tended to produce unbalanced, under-dimensioned people tailored to fit the modern economy" (quoted in Hicks and Holden, 1995). Since conventional educational institutions were tailored to meet the needs of our consumer industrial society, "it should not be surprising that our society's present direction aligns itself with programmes that inhibit human-earth relationships" (O'Sullivan, 1995).

THE CHALLENGES OF WORKING WITH ADULTS

There are obvious challenges that must be recognised and adapted to when working with adults in non-formal situations such as communities, that one does not necessarily encounter working with a group such as children or youths in an institutional setting. Adults often have less time in which to engage in workshops and seminars due to work and family responsibilities. Also, research shows they learn more when their experiences are acknowledged and used in the learning process. Therefore, since they view time differently than children and carry a variety of real-life experiences, their learning must be pragmatic, experience based, and quickly accomplished. In addition adults often (as do children) have a variety of learning styles that they bring to a workshop or group and their needs, concerns and viewpoints may be (and often are) completely divergent. For these reasons, an important principle underlying adult learning should be the inclusion of the adult learner in the identification of needs and objectives of any workshop or seminar (Camozzi, 1994; Burch, 1994).

Although environmental adult education is not yet as "comprehensively developed as it should be, there is already an important body of experience which can help inform the future development of policy and practice" (NIACE, 1993). Around the world, non-formal environmental education definitions, theories and practices for adults are being developed. They are approaches that are flexible in order to meet the needs of differing groups, members within it, and their specific situations (Camozzi, 1994). A key supporter of environmental adult education around the world is the Toronto-based International Council for Adult Education (ICAE).

The leaders in both the environmental movement and the field of education are adults, and therefore, they have the opportunity to teach and learn from one another. As the voters, decision-makers, government and community leaders, and consumers, adult non-formal education it would seem, has become an imperative.

LEARNING FOR ENVIRONMENTAL ACTION:
THE INTERNATIONAL COUNCIL FOR ADULT EDUCATION

The ICAE's primary philosophy of environmental adult education was most eloquently articulated by the late Victor Johnson, first Co-ordinator of the Environmental Education Network of the African Association for Literacy and Adult Education (AALAE). In his 1987 landmark book, 'Environmental Education Through Adult Education', the first book published in Africa (or anywhere in the world for that matter) to put forward activities and methodologies of environmental education solely for adults, Dr. Johnson described environmental adult education as a key tool to provide and develop knowledge, create understanding and awareness, build skills, a sense of commitment and responsibility, and ultimately stimulate individual and collective action. Environmental education is a tool to help people to work locally to improve their lives, to understand the international impacts and implications of what they do, and work together for political change.

In 1988, in response to the Belgrade and other conferences' calls for the strengthening and re-creating of environmental education, the ICAE began a process to create a world-wide environmental adult education programme. The first goals of the programme were to promote environmental education theory and practice within the educational work of popular-oriented social movements, strengthen and support the work of adult educators and activists around the world already engaged in environmental education, encourage the integration of a stronger educational dimension in the work of environmental non-governmental organisations (NGOs), solidarity and policy groups, and institutions, and stimulate global action around the environment, through education (Clover, 1995).

The idea began as simply to create an Environmental Education Programme. However, after much deliberation, it was decided that a key element in the struggle for a healthy planet was far more than just information sharing, it was the need to act. Therefore, it was unanimously agreed that the ICAE should establish an action-oriented global Programme and that it should be titled 'Learning for Environmental Action' (LEAP) because it linked the principle notion of on-going learning *vis-à-vis* the relationship between humans and the rest of nature with an action-oriented and political concept.

99

In January of 1990, during ICAE's Fourth World Assembly of Adult Education in Bangkok, Thailand, the Programme was solidified and the first co-ordinator, Moema Viezzer, a feminist educator / activist from Brazil was appointed. She identified the five key areas of the programme as: management participatory action research; information dissemination through the newsletter 'Pachamama' an Aymaran word meaning Mother Earth; media liaising, and, most importantly, mobilisation and action. In her presentation to the ICAE Executive Committee in 1991 in Gothenberg, Sweden, Moema articulated what she believed to be the premise of the LEAP and its next step:

"A programme that will offer an holistic approach which links environmental concerns to other social issues. The Programme will play an important role in strengthening the popular movements involved in environmental issues placing special emphasis on the 1992 UN Conference on Environmental and Development (UNCED)."

Participants of this meeting agreed with the suggestion of hosting a major event during the Earth Summit on environmental adult education. The mobilisation component of the Programme on a global scale was about to begin (Clover, 1995).

From June 1 - 4, 1992, under a tent in Flamenco Park in the heart of Rio de Janeiro, Brazil, the ICAE hosted the 'International Journey for Environmental Education'. The term 'Journey', used to describe the four-day seminar, was chosen because it best reflected the undertaking: an adventurous voyage into the developing field of environmental adult education. The key document developed during the Journey was the treaty entitled 'Education for Sustainable Societies and Global Responsibility'. This education treaty was one of 29 different citizen's treaties collectively entitled 'The Earth Charter' and was widely disseminated by the Global NGO Forum at the conclusion of the 'Earth Summit'. The 'Treaty' put forward a strong commitment to environmental education and its role in social and ecological transformation:

"We signatories, people from all parts of the globe, are devoted to protecting life on earth and recognise the central role of education in shaping values and social action. We commit ourselves to a process of education transformation aimed at involving ourselves, our communities and nations in creating equitable and sustainable societies. In so doing we seek to bring new hope to our small, troubled, but still beautiful planet."

The Treaty also recognises the fundamental right of humans to education, encourages action, declares that environmental education is not neutral but is value-based, and calls for education to stimulate solidarity, equality, respect for human rights, and provide an open climate of cultural interchange, dialogue and action. Moreover, it documents the importance of an education that is international or global in context because human activities, particularly those in the North, have such a resounding impact on the rest of the world.

The principles espoused in the Treaty have become a basis for teaching adults in many parts of the world. They form curriculum material for adult literacy classes in Latin America, training workshops in Asia and popular theatre skits in Africa. The 'Treaty', the 'Belgrade Charter', and other important international statements have provided a framework to develop definitions, theories and methodologies of environmental adult education around the world.

DEFINITIONS AND THEORIES OF ENVIRONMENTAL EDUCATION

"Scientific understanding is not going to change our habits or give us   the political will to change our life. Even the hard facts which tell us we should not do this or that do not actually persuade people as much as spiritual experience can. You have to reach the hearts of the people" (Jaquit Chuhan, quoted in Deri & Cooper, 1993).

The field of environmental education is having to take on new intellectual, emotional and spiritual energy "to keep pace with the devastating urgency of environmental problems" (Corcoran and Sievers, 1994). Changes in how humans view the world today have meant the re-conceptualisation of environmental education in order to give "greater prominence to the root social, political and economic causes of the environmental situation" (Tilbury, 1995). Guiding principles have become greater and more varied. The educational framework appropriate for today's world must be more visionary and transformative, going beyond the conventional educational outlooks that have been cultivated for several centuries in the name of conquest and industrialisation (O'Sullivan, 1995).

Environmental education, like conventional education, has itself been widely critiqued. A few of the critiques have focused on the fact that it has been primarily apolitical, informed by reductionist empirical science, has focused on natural resource use and/or management as a key goal, and been geared towards recreational pursuits only (Burch, 1994; Hammerman and Hammerman, 1973; Tilbury, 1995; Rowe, 1990).

Apart from these weaknesses, many environmental educators have developed an educational practice that has the ability to help people connect with the rest of nature through sensory perception and emotion which arises from the idea that understanding can flow from feelings and intuitions as much as from scientific knowledge. Moreover, their practice has been a way to share an understanding of and ability to communicate, the inter-connectedness of everything and intrinsic value of nature by using nature as the teacher and site of learning. Educating out-of-doors or 'outdoor education' is a concept and practice that recognises and respects the vital function, beauty, rights, and pedagogical importance of the natural world. These practices of teaching and learning were developed before the current focus on the environmental crisis, and is gaining recognition in many parts of the world such as Chile, Fiji, Uganda, Hong Kong and Canada to name but a few (for examples see Vio Grossi, 1996; Pachamama, 1995; Burch, 1995; Lechte, 1993).

Environmental education today has been challenged to consider the environment in its totality, natural and built, technological and social, and view itself as a continuous life-long process that takes place both inside and outside the classroom. It must assist people to make global and local links, and understand the importance of valuing the inter-relatedness of the web of life, and promote concern for all life on the planet (Tilbury, 1995). Environmental education today must promote the development of critical thinking about environmental problems and become more politically literate about the ideologies and systems at work that are contrary to the promotion of a healthy social and natural environment. It must learn and draw from the knowledge of indigenous people, those who are most often closest to the land.

The ICAE calls for environmental adult education world-wide to support new models of environmental adult education that help people in the North to recognise and work towards the reduction of consumption and dominance; acknowledge the complex relationships between the environmental crisis and poverty, human rights, health, and the occurrence of 'not so natural' disasters such as flooding, desertification, and so on; recognise that earlier divisions between 'North' and 'South' or between 'developed' and 'developing' mask how people of colour, indigenous peoples, the differently abled, and persons of other non-dominant social identities are disproportionately affected by toxic waste, polluted waters and other biospheric contaminants; support a process of transformative learning which helps people to re-connect with the natural not simply through the mind but also spiritually and through the senses of sight, smell, taste, touch and hearing; use learning as a tool to build sustainable societies and encourage participation; and build a vision in which the contributions of women, men, persons of colour, indigenous persons, the young, the old and differently abled are equally respected.

Environmental education for adults has to be more participatory, action-oriented, experiential and begin from where people are, in their homes and communities. To borrow from feminist theory, it must be with, for and about the environment. Environmental education for adults around the world has begun to incorporate these and many of the guiding principles presented in the governmental and non-governmental documents mentioned previously.

Matthias Finger, a leading European adult educator, in his research report 'Environmental Adult Learning in Switzerland' (1992), notes that education cannot become merely "a substitute for behavioural change or action. More environmental information and knowledge will not necessarily translate into corresponding behaviour." He recommends therefore, that "social and collective action be an integral part of any continuing education activity" and that educative experiences take place in the natural environment. He concludes his report by stating that:

"[Environmental education] programmes should provide simultaneously, nature experiences and experiences of environmental activism, deal with fear and anxiety, and transmit relevant environmental knowledge."

Effective environmental education will need to include a socio-political analysis of humanity's response to itself and oppressions such as humans over nature, men over women, rich over poor, whites over people of colour, engage adults in activities that re-connect them with the rest of nature, by using nature as the teacher and site of learning. Most importantly, it must be experiential and promote action both individually and collectively.

Like other forms of liberatory education, environmental adult education must be political in its content and have as its goal social and ecological change.

THE PRACTICE OF ENVIRONMENTAL ADULT EDUCATION

"Theory - the seeing of patterns, showing the forest as well as the trees. Theory can be a dew that rises from the earth and collects in the rain cloud and returns to earth over and over. But if it doesn't smell of the earth, it isn't good for the earth." (Adrienne Rich, 1993)

There are numerous non-formal environmental adult education practices taking place around the world today. They are more political, based on people's experiences, participatory and experiential, creative, action-oriented and use the rest of nature as the teacher and site of learning.

In Hong Kong, two-day community workshops provide an opportunity for concerned adults and community-based environmental educators to discuss local and global environmental issues and to share their educational and community development experiences. The workshops begin with a participatory discussion of the environment *vis-à-vis* the problems and role of people as contributing factors to the environment. The outcome of the workshop is a strengthening of participants to continue to work for the conservation of the environment and a greater awareness of global, regional and local environmental issues. A key component of the workshops are field visits.

In one particular workshop two different areas were visited. The first was the River Indus near a highly industrial area. Participants viewed first-hand the problem of river pollution and its associated impacts on the environment. The second was a visit to Mai Po Marshes Nature Reserve to learn about the various kinds of endangered birds such as Black Faced Spoon Bill and rare plants such as mangroves. Participants also attended lectures, slide show presentations and a film on the conservation of biological diversity in Hong Kong. The slide presentation shared interesting information on local flora and fauna. Local legislation on the preservation of wild species and habitats was addressed briefly during the lecture. At the conclusion of the slide presentation, a quiz was given to participants to test their knowledge on local biological diversity and the 'ecology game' was used to learn about various components of the ecosystem (Pachamama, 1995).

The 'Navdanya project' in India is a community-based group that uses celebration of diversity and resistance as ways of uniting education and action in a popular form. Literally, 'Navdanya' means 'nine seeds' and metaphorically it represents balance based on diversity at every level, from the cosmic domain to the community, from the ecology of the earth to the ecology of the body. Education sessions begin with the experiences of the individual farmers who examine the legacies of the Green Revolution such as what monoculture farming has done to their communities, what they have lost and what they hope to gain by moving back to organic farming and crop diversification. Community festivals centred around seeds are used as ways to bring communities together and learn from each through celebration (Ramprasad, 1993).

In addition to workshops, festivals and celebrations, large demonstrations have taken place across India against genetic engineering, the green revolution and World Bank / IMF projects. Farmers have come out by the thousands to protest against government decisions to allow the patenting of the neem tree and the building of hydro electric dams by the World Bank (Shiva, 1989).

The 'Ugandan Multi-Purpose Training And Employment Association' (META) organises training workshops on environmental education with community members. The goals of the workshops are to strengthen environmental awareness through adult environmental education, develop an adult environmental education curriculum for the communities that is drawn from the 'Environmental Education for Sustainable Societies and Global Responsibility Treaty' developed in Rio de Janeiro; encourage, initiate and strengthen activities within the community that are environmentally sound, and create an active networking strategy through the development of training, educational practices, techniques, monitoring and evaluation skills. The workshops are participatory and experiential and incorporate diverse and creative activities such as presentations, brainstorming and small group discussions, role-playing, storytelling, drama, music and field visits.

The key mandate of the 'Environmental Popular and Adult Education Programme' (EPAE) of the North American Alliance for Popular and Adult Education (NAAPAE) is to work locally in North America and globally through the ICAE to introduce the practices and philosophies of environmental popular education into the environmental movement, in order to help strengthen the educational component of the work and raise awareness within the global adult education movement of the link between ecological and social issues.

The major project of the EPAE within Canada is to develop a practice of environmental education that is experiential and participatory rather than expert-driven and information heavy, raises political awareness and promotes action on the local and at other levels, and re-connects adults with the rest of nature. This is being done through the organisation of workshops with community groups across Canada.

The primary purposes of all the workshops, although they vary according to specific needs within communities, are to understand the principles of adult, popular, feminist, and environmental-popular education and how they provide vehicles for action; to analyse relations of power in the context of environmental justice; to connect with the rest of nature through sensory perception and emotions; to provide the opportunity for people from various sectors of society (health, environment, economic, community, etc.) to engage effectively and creatively with one another; and to share and learn about educational practices and strategies in Canada and around the world.

The workshops begin with activities that create a relaxed atmosphere by using humour, having people move physically about and describe themselves in terms of an animal, plant or any other element of nature. Another key component of each workshop is the 'Fishpond', an activity in which facilitators introduce the principles of popular, adult, feminist and environmental-popular education and how they can be used for creative learning, action, and developing stronger working relationships amongst community members. The workshops use a variety of activities to understand various relations of power, nationally, globally and within community groups; they discuss the marginalisation of peoples and the rest of nature that often takes place in attempting to create a healthy community, and includes field visits to local department stores to analyse the politics and impact of consumerism on their community and the world. Outdoor activities are used to re-connect participants with the rest of nature through the senses of touch, smell, hearing, and sight. Adults identify the positive and negative aspects of the built and natural worlds. As a central component of environmental popular education is action, workshops conclude with a session that brings participants together to strategise around and share educational practices that are effective and creative when working with adults.

WOMEN AND ENVIRONMENTAL EDUCATION

Research shows that women are the first environmental educators, connecting with their children and passing along an understanding of the natural processes which take place around them (Dankelman, 1988; Rodda, 1991). Around the world they have become increasingly more adept educators, communicators and information specialists *vis-à-vis* the environment (Rodda, 1991). Through their education practices, women have become active agents of change in society. The teaching methodologies and practices they use to foster environmental action are inclusive and sensitive to gender. Women's education methodologies aim at empowering those with little or no power and enabling people to see that they do have the power to work towards social and ecological change the world (Ba N'Diaye, 1993; Rodda, 1991; Dankelman & Davidson, 1988).

Women living in the interior of larger Pacific islands like Fiji depend on biological resources for health-care, medicines, food, fuel and craft materials for their families' and communities' subsistence, and cultural and economic needs. In light of this, an environmental education community workshop was organised in Fiji to take women into the rainforests to explore their relationships with the ecosystem through experiential exercises. For five days, women lived with Nadovu villagers, experiencing their dependence on the productive rainforest through role-playing, identifying uses of trees and plants, hikes, songs, and theatre performances. The workshop ended with a visit to the local sawmill and logging operation where the women observed oil leaks, soil erosion and sawdust contamination. Through these activities, women were able to increase their awareness of the value and fragility of biological diversity and see the impact of human activity on the natural world around them. The women emerged from the forest "with a firm commitment to environmental management and to protecting and sustaining resources" (Pachamama, 1995).

Another example is the Atlantic Women's Fishnet of Canada, a network of women in the Atlantic provinces concerned with the current fishery and environmental crisis in their coastal communities. This group uses popular theatre with men and women in communities as a way to educate about the ecological disaster, brought about by corporate drag-netting, that resulted in the death of the codfish and, therefore, the fishing industry that supported a vast majority of the population. Through active community participation in theatre and role-playing and information exchange through a newsletter, they also analyse the exclusion of women's voices from government research and planning in regard to the fishery and from training programmes and financial assistance (Fishnet, 1994).

In Bolivia, women use songs, popular theatre, video, and other media are used as educational tools for working in communities with adults around environmental issues (Lechte, 1993). In India, women hug trees to stop them from being felled, and throw themselves in front of bulldozers to prevent the clearing of their land for industrial use (Shiva, 1989). A great deal of learning takes place, particularly around human rights, the rights of nature, and political manipulation, both locally and globally, through taking direct action. Women concerned about toxic materials in Sudan used workshops and the mass media for education. "They lobbied the government to ban dangerous pesticides and pharmaceuticals such as skin lightening products used by Sudanese women which contained highly dangerous heavy metals" (Lechte, 1993). And the list goes on and on.

Environmental adult education methodologies within communities show how the knowledge of others can become a basis to create new knowledge and a new vision of the world. They are processes of learning that begin with the experiences of the learner, involve a high degree of participation, are action-oriented, and work towards democracy and social justice. The training projects, workshops and political action measures provide a wealth of learning which often leads to a transformation in thinking and acting for many adults who would otherwise not be exposed to such positive and challenging activities.

CONCLUSION

"...there being an important link between the 'quality of the environment' and the 'quality of education'" (Keith Wheeler, 1992).

There is truly a challenge facing educators today. This is, according to Hall and O'Sullivan (1994), that

"all education, up to our present moment, has never countenanced the possibility of planetary destruction. This viability of planetary existence has never been an issue for educators because it was not, until now, part of our cultural understanding."

In light of this challenge, including a focus on the environment within the field of adult education is timely and relevant. Because education can be an effective tool to secure change and achieve a more just and sustainable society (Orr, 1992), it is necessary that it focus on all people and at all levels as noted so well in 'The Belgrade Charter'. Smyth (1995) notes that "more attention is now being paid to community education as a vehicle of environmental education" as a more holistic way of dealing with the environmental crisis.

Drawing strength and inspiration from 'The Belgrade Charter' and other international policy statements, adult educators around the world are developing environmental education for adults who are not part of pedagogical institutions but who "outfit each youth with cultural belief-spectacles that direct and colour a particular world view [and] once formed, it becomes both self-evident and 'right' for those born to it" (Rowe, 1990). The major influence adults have on children and their decision-making power (although some have more than others) are key reasons why developing critical environmental education practices are so important.

Non-formal environmental adult education implies a break with current education practices that are top-down, expert-driven, and maintain, stabilise and legitimise the contemporary industrial social order. It puts forward methods that are creative, dynamic, flexible, and highly participatory.

These practices include a socio-political analysis of modern day complexities, beginning in people's homes, drawing on their experiences to develop new knowledge and take action. Often, they understand the value of being with and using the rest of nature as a place where feelings and creativity can be stimulated and knowledge can be gained as well as the need to take direct political action. Environmental adult education practices are ones out of which meaning constantly emerges because

"different influences predominate at different times of life and in different circumstances: collectively, they are a sustained and lifelong learning experience" (Smyth, 1995).

REFERENCES

Ba N'Diaye, S. and Moussoulimou K. (1993) 'Theory and Practice of Feminist Popular Education.' In *Jol. Voices Rising.* No.1.

Burch, M. (1994) 'Adult Environmental Education: A Regional Report for North America'. Paper prepared for the *International Council for Adult Education* and presented in Curacao, Netherlands Antilles, May 12-28.

Burch, M (1995) *Simplicity.* New Society Publishers, Gabriola Island.

Camozzi, A. (1994) *Adult Environmental Education, A Workbook to Move f from Words to Action.* EcoLogic and Associates, Antigonish.

Clover, D. (1995) 'Learning for Environmental Action: Building International Consensus'. In Cassara B. (Ed.) *Adult Education Through World Collaboration.* Krieger Publishing Company, Malabar.

Corcoran, P. and Sievers E. (1994) 'Reconceptualizing Environmental Education: Five Possibilities'. *In Journal of Environmental Education.* Vol. 25 (1) pp. 4 - 8.

Dankelman, I. and Davidson, J. (Eds.) (1988) *Women and Environment in the Third World, Alliance for the Future.* Earthscan Publications Ltd. in association with The International Union for Conservation of Nature and Natural Resources (IUCN), London.

Deri, A. and Cooper G. (1993) *Environmental Education - An Active Approach.* Regional Environmental Centre for Central and Eastern Europe, Hungary.

Finger, M. (1992) '*Environmental Adult Learning in Switzerland'.* Final R Report to the Swiss National Science Foundation.

Fishnet (1994) *Atlantic Women's Fishnet Newsletter,* Vol.1(1) Winter 1994/95.

Hall & O'Sullivan (1994) 'Transformative Learning: Contexts and Practices'. In Transformative Learning Centre *Awakening Sleepy Knowledge: Transformative Learning in Action.* Toronto, OISE Press.

Hammerman, D. and Hammerman W. (1973) *Teaching in the Outdoors.* Burgess Publishing Company, Minneapolis.

Hicks, D. and Holden C. (1995) 'Exploring the Future: a missing dimension in environmental education.' *In Environmental Education Research.* Vol. 1 (2) pp. 185-193.

Ibikunle-Johnson, V. and Rugumayo E.(1987) *Environmental Education Through Adult Education.* African Association for Literacy and Adult Education, Nairobi.

Lechte, R. (1993) 'Women as Educators for Primary Environmental Care'. Paper written for the *May 1993 Seminar on Environmental Education for Women in Asia.* Taiwan.

NIACE (1993) 'Learning for the Future'. Special issue on adult learning and t the environment, In *Adults Learning,* Vol. 4(8).

Orr, D. (1992) *Ecological Literacy: Education and the Transition to a Postmodern World.* State University of New York Press, Albany.

O'Sullivan, E. (1995) *Education and the Dilemmas of Modernism: Toward an Ecozoic Vision, The Dream Drives the Action.* OISE Press, Toronto.

Pachamama (1995) Numbers 1 and 2, ICAE, Toronto.

Ramprasad, V. (1994) *Navdanya, New Delhi: Research Foundation for Science, Technology and Natural Resource Policy.*

Rich, A. (1993) 'Notes Towards a Politics of Location' In Gaard G.(Ed.) *EcoFeminism: Women, Animals, Nature.* Temple University Press, Philadelphia.

Rodda, A. (1991) *Women and the Environment.* Zed Books Ltd., London.

Rowe, S. (1990) *Home Place.* New West Publishers Ltd., Edmonton.

Shiva, V. (1989) *Staying Alive.* Zed Books Ltd., London.

Smyth, J. C. (1995) 'Environment and Education: a view of a changing scene' In *Environmental Education Research.* Vol. 1 No.1, pp. 3-20.

Tilbury, D. (1994) 'The International Development of Environment Education' In *Environmental Education and Information.* Vol. 13 (1) pp. 1-20.

Tilbury, D. (1995) 'Environmental Education for Sustainability: defining the new focus of environmental education in the 1990s'. *In Environmental Education Research,* Vol. 1 (2) pp.195-212.

UNESCO-UNEP (1976) *Connect* Vol. 1(1) January.

Viezzer, M. (1986) 'Learning For Environmental Action'. In *Convergence.* Special Issue Vol. 24 (2) pp. 3 - 8.

Vio, G. (1996) 'New Paradigms, the Ecological Society and Education'. In *Convergence*. Vol. 29 (1&2).

Wheeler, K. (1992) 'International Environmental Education: A Historical Perspective'. *Environmental Education and Information.* Vol. 4 (2).

Chapter 7

BUILDING BRIDGES: TRADITIONAL ENVIRONMENTAL KNOWLEDGE AND ENVIRONMENTAL EDUCATION IN TANZANIAN SECONDARY SCHOOLS

Christian Da Silva

International Development Research Centre
PO Box 8500
Ottawa K1G 3H9
Canada

INTRODUCTION

Interest in environmental education has increased in the last few years, due in part to the high profile the subject received during the 1992 'Earth Summit' in Brazil. In an atmosphere of high expectations and hopes for concrete solutions to gripping environmental problems, environmental education re-emerged as a relatively innocuous and politically palatable catch-all.

I say 're-emerged' because environmental education has had its day on the world stage several times before[1]. Beginning with the 1972 United Nations conference in Stockholm, a series of international meetings have, to varying degrees, dealt with the development, promotion, and legitimisation of environmental education (e.g. Stockholm, 1972, Belgrade, 1975, and Tbilisi, 1980). The 1975 'Belgrade Charter' (UNEP, 1976) set out a global framework for environmental education that proposed to "develop a world population...aware of and concerned about the environment and its associated problems, and which has the knowledge, skills, attitudes, motivation and commitment to work individually and collectively toward solutions of current problems and the prevention of new ones" (UNEP, 1976). Belgrade inspired UNESCO and UNEP to work together to create the International Environmental Education Programme (IEEP).

---

[1] For an extensive history of environmental education in the United Nations, see Meadows (1989:61-2).

Why all the attention afforded to this subject? Environmental education is a term which conjures up visions of a panacea; a progressive subject which teaches values and practices with the potential to reverse environmental degradation and promote harmonious and sustainable living. While environmental education cannot single-handedly lead to political, social or economic reform to the well-entrenched power structures of schools or states, it may have the potential to engender in people, particularly younger generations, an awareness and critical analysis of environmental problems, including their systemic root causes and possible solutions. It could be argued that the potential for environmental education to be an agent of change is even greater if the environmental concerns, beliefs, and ideas of local populations are recorded, acknowledged, and incorporated into environmental education initiatives targeted at those same populations or at people who will play important roles in the civil institutions that impact on local people.

Data will be presented in this chapter to illustrate the considerable gap which currently exists between local and school-based knowledge systems. I will argue that environmental education can and should be positioned as an alternative 'wedge' within the Tanzanian secondary education system in an effort to highlight environmental issues, degradation, and resource management strategies, through a better understanding of the traditional environmental knowledge base found at the village and community levels. I will argue that environmental education can serve as a counter balance of community-based relevance to the dominant educational system rooted in the political economy of modernisation and global capitalism. Finally, I will discuss a practical methodology for discovering and documenting local environmental knowledge (LEK) for use in environmental education programs.

ENVIRONMENTAL EDUCATION; WHAT, WHY AND FOR WHOM?

Taking a 'narrow' view of environmental education as practised in formal secondary school situations, it must be recognised that environmental education is just another branch of the education system, the main purpose of which is "the replication of society and the maintenance of a stable social order" (Fitzgerald, 1990). One must not be seduced by the ameliorative connotation of the term 'environmental education', for, like most developments in education, content and philosophy are "determined first and foremost by a mutually reinforcing mixture of power and money" (Fitzgerald, 1990), while the influence from moral and ethical forces is limited. On the other hand, environmental education can be defined; personally, locally, even nationally, and it can serve as a 'starting point' in achieving environmental sustainability. In establishing a definition of environmental education in the formal sector, I offer the following outline by Johnson (1992):

"...A well designed environmental education programme should provide students of whatever level with learning experiences that will enable them to develop the following: a) Intimate knowledge of ecological parameters, their nature, characteristics as well as means of qualitative and quantitative assessment; b) the ability to predict likely changes in the environment; c) the ability to propose suitable adaptive responses to changes in the environment and to appreciate and assess the limitations of such responses; d) ability to influence the power structure in communities in favour of making effective responses to environmental changes; e) critical thinking".

In many developing countries such as Tanzania, few people have the opportunity to attend secondary school. Those who do invariably assume key positions within society upon completion, working as extension workers, educators, and bureaucrats. This being the case, secondary schools have a crucial role to play in influencing future 'change agents'. Recognising that environmental education is an interdisciplinary activity focused on problem solving and critical analysis, the content of environmental education programmes should reflect the values, concerns, and language of the communities that will be impacted, managed, or otherwise influenced by new graduates of the school system. Secondary school graduates must be willing to listen and able to dialogue about environmental concerns and possible solutions in a way that is cogent with the views of local people - farmers, peasants, artisans etc. Ideally, secondary education should include a serious examination of LEK, its methods of storage, transmission, and survival.

Thai environmentalist, Witoon Permponngsacharoen (1992), suggests we ask basic questions with respect to environmental education and local communities such as; what are the causes of environmental problems? Where do poor people fit in? How are they affected by environmental degradation? What do *they* [my emphasis] think about these problems? Conducting formal environmental education from a bottom-up perspective, therefore, necessitates more active and participatory strategies which, in turn, will lead to a more holistic understanding of issues.

People must be able to exchange information and ideas from what Gallopin (1991) calls the same 'socio-ecological' perspective. Ideas that incorporate local ways of knowing, both in the description and amelioration of a problem will be more effective and more readily accepted. Since environmental education acts on the values, attitudes and behaviours of learners, it is incumbent on the education system to ensure that environmental education reflects the values, attitudes and behaviours of the communities in which schools are situated.

DEFINING LOCAL AND WESTERN KNOWLEDGE SYSTEMS

There is considerable mixing of terminology in the literature dealing with non-western knowledge systems. Traditional ecological or environmental knowledge (TEK) and indigenous knowledge (IK) are two examples of more commonly-used expressions. The differences are mostly semantic, though some definition is warranted. Strictly speaking, traditional "...refers to cultural continuity transmitted in the form of social attitudes, beliefs, principles and conventions of behaviour and practice derived from historical experience" (Berkes, 1993). Societies, however, continuously react and adapt to exogenous factors. The word 'traditional' implies a more static experience which minimises the usefulness of the word 'traditional' in describing a dynamic, adaptive knowledge system.

Similarly, ecological knowledge can be too narrowly confined to the disciplinary boundaries of biology and botany rather than to the socio-ecological interactions at work in an environment where people live. A preferable definition for this discussion is a term which acknowledges contemporary influences on indigenous knowledge systems and also encapsulates the socio-ecological dynamic; something more akin to what Levi-strauss has called the 'science du concret' or native knowledge of the natural milieu (Berkes, 1993). Therefore, I use the term Local Environmental Knowledge (LEK) throughout the paper. Figure 1 presents some of the characteristic differences between Local and Western knowledge.

At the risk of oversimplifying the case, the success or failure of knowledge systems closely mirrors that of its adherents. The current dominance of Western scientific knowledge over that of African pastoralists, for example, is a reflection of the relative technological accomplishments of these two groups of people respectively. Western science brings with it remarkable achievements in science and technology over a relatively short period of time. Indigenous knowledge systems, on the other hand, have endured for millennia, but have not demonstrated the same dramatic technological leaps. It is partly for this reason that local or indigenous knowledge systems have come to be regarded as inferior.

COMPARISON	LOCAL KNOWLEDGE	WESTERN SCIENTIFIC KNOWLEDGE
Relationship	Subordinate	Dominant
Dominant mode of thinking	Intuitive (holistic)	Analytical (reductionist)
Communication	Oral (storytelling, subjective experiential)	Literate / didactic (academic, objective, positivist)
Data Creation	Slow/inconclusive	Fast / selective
Prediction	Short-term cycles (recognise the onset of long-term cycles)	Short-term linear (poor long-term analysis)
Explanation	Spiritual (the inexplicable)	Scientific inquiry (hypothesis, laws)
Biological Classification	Ecological (inconclusive, internally differentiating)	Genetic and hierarchical (differentiating)

Adapted from Lalonde (1993) and first cited in Wolfe et al. (1991).

Figure 1: A comparative framework of local and Western knowledge

Like indigenous knowledge systems, Western knowledge is also cumulative and experiential. It assumes, however, a dominant, widely accepted form of rationality and it is this dominant rationality, and its related successes, that makes other ways of knowing appear modest in comparison.

Western knowledge is easily recognised by the formal status accorded its practitioners and institutions. The system of secondary education in Tanzania reveals remnants of the British system; academically focused, strongly based on the absorption of theory and concepts, and reliant entirely on formal examinations. This leads, as it does in North America, to diplomas, degrees etc. This system of accreditation is further validated by government and quasi-government institutions which seek graduates of the education system, rewarding those who have gone the furthest. Certainly, societies need skills, such as literacy and numeracy, which secondary schools provide. However, there remains the question of whether or not these same schools can also be made to be more inclusive of community concerns and historically marginalised knowledge.

PHILOSOPHICAL UNDERPINNINGS OF TANZANIAN SECONDARY EDUCATION

Julius Nyerere, the man who led Tanzania to independence and ruled the country until the mid-eighties, had a deep interest in education and educational policy throughout his Presidency. He too recognised the integrity, value and complexity of indigenous education. He writes in Fafunwa *et al.*, (1982):

"The fact that pre-colonial Africa did not have 'schools'...did not mean that children were not educated....They learned the tribal   history, and the tribe's relationship with other tribes and with the   spirits, by listening to the stories of their elders. Through these means, and by the custom of sharing to which young people were taught to conform, the values of the society were transmitted."

The influence of German and, later, British rule on Tanzanian education was profound. It was this system, based on racial segregation between African and European and a world view born of the metropole, that Nyerere inherited in 1964. Of this inheritance, Nyerere wrote:

"Colonial education in this country was...not transmitting the values and knowledge of Tanzanian society from one generation to the next: it was a deliberate attempt to change those values and to replace traditional knowledge by the knowledge from a different society"   (Fafunwa *et al.*, 1982).
Nyerere attempted to reform the educational system to make it more relevant to Tanzanians[2]. He articulated his views directly in the policy document entitled 'Education for Self-Reliance' (Nyerere, 1967). However, looking back in the early 1980s, Nyerere still found the fundamental faults in the educational

---

[2] One of the most notable recommendations of the 1967 policy was the establishment of community schools. Buchert (1992) describes the community schools as an attempt to recreate, for the modern context, a school based on the pre-colonial African way of transmitting mental and manual skills to children. Buchert studied two community schools and found that, while the original aim of promoting cooperation between school and community was generally achieved, constant changes to the objectives and personnel involved in the pilot schools diluted early successes. Buchert questions whether the more far reaching purpose of promoting joint school-community problem solving was ever achieved and cites structural reasons for his doubts. For example, the concept of the community school was constantly being changed at the National level, while implementation was poorly supervised and left entirely to the local communities. No comprehensive assessments or evaluations were utilized to monitor the community schools initiatives and it was officially abandoned in 1982.

system which existed at independence. He identified three persistent weaknesses:

1. Schools provide an elitist education designed to meet the interests and needs of a very small portion of those who enter the school system.
2. Tanzania's education [system], divorces its participants from the society it is supposed to be preparing them for. This is particularly true of secondary schools, which are almost entirely boarding schools.
3. The educational system promotes the idea that all knowledge which is worthwhile is acquired from books or from 'educated people' - meaning those who have been through a formal education. The knowledge and wisdom of old people is despised, and they themselves are regarded as being ignorant and of no account (Fafunwa, 1982).

Although Tanzania's educational system has continued to evolve, policy-makers today still view the secondary system as an instrument of modernisation and a source of skilled workers for a science and technology based economy. Tanzania's Integrated Education and Training Policy (1993) states that:

"The acquisition of scientific and technological skills by the majority of school graduates as the country moves towards the 21st century... will ultimately have a direct bearing on the social and economic development of Tanzania. Specifically, it is assumed that by imparting scientific knowledge to the vast majority of school graduates, the country will be creating an adequate reservoir of manpower that will exploit substantially the abundant natural deposits which it is endowed with. In turn, the exploitation of the natural resources, should have positive implications for raising the general standards of living for the majority of the people."

This policy statement would indicate that the weaknesses identified by Nyerere more than a decade ago are still present and more profound today. Is the recent interest in environmental education a sign of a new balance emerging within the secondary school curriculum or merely another trend characterised by half-hearted reform and good intentions?

SECONDARY SCHOOLS AND ENVIRONMENTAL EDUCATION

Various interests compete for time in Tanzania's school curriculum. The education system is a vehicle for the dissemination of social, political and cultural information, all of which are subject to changing priorities and cyclical popularity. Teachers are increasingly expected to explore issues beyond their own formal academic training and comfort level. Environmental education is no different among competing interests and, as awareness of environmental degradation has increased, so too have the calls for time on the school agenda, either as a designated course or through an integrative or infused curriculum approach.

There is an increased effort on the part of NGOs[3], individuals within the Ministry of Education and the Curriculum Development Institute, and the University of Dar es Salaam, as well as teacher training colleges, to improve environmental education programming in secondary schools. Environmental education forms a major part of Tanzania's National Conservation Strategy as outlined by the National Environment Management Council. NGOs, such as the World Wildlife Fund, are committing substantial resources to provide environmental education training to teachers and policy makers. Numerous documents collected during the course of this research indicate a strong and renewed emphasis emanating from the University, NGOs and various Ministry officials for improving environmental education design and delivery[4].

Although efforts have been made to introduce environmental education into the curriculum, some obstacles exist. For example, the current structure[5] of secondary schools themselves has proven to be a hindrance in the development and promotion of environmental education. One official from Tanzania's National Environment Management Council (NEMC) characterised the situation for introducing environmental education in secondary schools this way:

---

[3] A mail survey was conducted during my 7 months in Tanzania which was targeted specifically at 28 Tanzanian NGOs. Among other things, the survey asked whether the organisation considered environmental or conservation education to be part of their mandate. Eleven NGOs responded - a respectable return rate for a nascent volunteer sector - and all indicated that environmental education was part of their work.

[4] A number of unpublished papers and internal documents support this claim such as Osaki *et al.*, 1994; Bakobi, 1992.

[5] Structure in this context refers to the boarding school arrangement which disconnects students from the surrounding community, the paramount importance placed on national examinations as described earlier, and the government focus on a modernization paradigm in the educational system as articulated in recent policy documents.

"Existing concepts in the curricula are geared towards equipping a student to make his/her own living. Studies are pursued in order to pass examinations, obtain academic certificates and hence earn a living. There are inadequate efforts to make education something interesting and worth pursuing even without a certificate, (for example, the development of studies on environmental conservation as an extra-curricular activity)" (NEMC, 1994).

Other problems exist as well. Most attempts at environmental education in Tanzania have sought to introduce environmental issues within existing disciplines (the infusion approach), rather than to create new interdisciplinary environmental education courses. Arguing against 'infusion', most environmental education experts in Tanzania criticise the compartmentalisation of subject matter in the school system generally, and point out the inappropriateness of such an approach in light of the interdisciplinary and holistic nature of environmental problems and solutions (Bakobi, 1992; Osaki *et al.*, 1994). The environmental education found in Geography, Biology and Civics curricula, for example, lacks coherence, integration, and practicality. Current environmental education practices make very little use of the actual environment as a teaching resource. Environmental concepts are borrowed from European or North American sources, rendering some if not all of the material as irrelevant. For example, the importance of the 'greenhouse effect' concept - while indisputably relevant and widely accepted in North American classrooms - was incomprehensible for many Tanzanian students. Why? The analogy of a greenhouse used to explain the phenomena, is one with which Tanzanian students have no connection or direct experience. The divorce between pedagogy, curricula and culture diminishes the effectiveness of environmental education programs.

LOCAL KNOWLEDGE AS ENVIRONMENTAL EDUCATION CONTENT

Is there evidence that by recognising and trying to utilise LEK in school curriculum communities might move further along the continuum toward environmental sustainability? There are clear patterns of survival which suggest the existence and use of highly specialised and valuable systems of local knowledge or wisdom. Permponngsacharoen (1992), says:

"...environmental education that builds on local knowledge in a way that is both appropriate and effective for solving current problems must start from the recognition that local people possess wisdom about dealing with their environment. Otherwise, how could they have survived for so many centuries, or in some cases, millennia? As such, the educator is obliged to study existing knowledge in order to understand how problems are dealt with traditionally."

Admittedly, many traditional practices characterised as local environmental knowledge are no longer sustainable given increasing populations, dwindling resources, and changing patterns of human settlement, production, and consumption. We must be careful not to automatically assume that what is *traditional* is necessarily sustainable or even desirable. For example, slash and burn agricultural methods, sustainable and rational in a pastoralist culture, may be catastrophic in a sedentary production system. Weaknesses in LEK need to be recognised, but local knowledge systems generally, should not be automatically discounted or marginalised because of their unconventionality.

Unfortunately, development processes everywhere have often run roughshod over indigenous practices or beliefs. As Redclift (1987) points out, "in the course of development, indigenous environmental knowledge is often lost because it becomes less relevant to the new situation and because it is systematically devalued by the process of specialisation around competitive production for the market."

Secondary education systems have also specialised, rushing to provide the modern tools needed for market economies, industrial development, and technological innovation at the expense of intuitive ways of knowing. These economically-driven modalities of education are reflective of rational, scientific ways of knowing, culturally laden with the values and beliefs of Western societies more typically detached from, rather than immersed in, the natural environment.

Over time, the knowledge that communities have accumulated regarding survival in their environment has eroded as have the socio-political systems that acted as the conduit for LEK. Dominant approaches to resource utilisation, for example, are approaches which Redclift calls 'managerialist' (1987). Not only are managerialist approaches to resource utilisation out of step with local environmental taxonomies and strategies, so too are they inappropriately conveyed.

"...If we want to know how ecological practices can be designed which are more compatible with social systems, we need to embrace the epistemologies of indigenous people, including their ways of organising their knowledge of their environment. However, as Norgaard argues, traditional knowledge is location specific and only arrived at "through a unique co-evolution between specific social and ecological systems". This knowledge is not easy to incorporate into 'scientific' knowledge since experiential learning requires an evolutionary rationale, and one which is different from that of bureaucratically managed institutions" (Redclift, 1987).

It is important for educators to remember that just as there are different ways of learning, so too are there unique ways of knowing. Meadows (1989) writes; "rationality is not the only human gift and it is not the only way of knowing something...a good farmer or forester knows much that is wordless about the soils, the plants, the animals, the trees - that understanding comes from familiarity, instinct, intuition, a sort of resonance with or tuning-in-to or empathetic observation of the living world". The reductionism of Western science is only occasionally questioned, and even that has come about only recently as 'scientists' seek holistic solutions to environmental and other problems.

Development efforts of the last several decades have largely ignored the cultural dimension of the societies being 'assisted', viewing these factors as a nuisance or impediment to modernisation. Subsequent loss of cultural identity and self-respect has been attributed to such mechanistic development thinking. Ironically, because the cultural dimension has been ignored for so long, development theorists and practitioners are unsure as to how to re-incorporate this dimension into their work. Development rooted in a modernisation paradigm has caused disruptive cultural change. There is a displacement of traditional values by modern ones.

Arguably, the most tangible legacy of Western influence which persists in African society is that of formal education. African educational systems have grown paradoxical. Education is alternately criticised and praised, first as the culprit responsible for the irreversible acculturation process that has been forced on Africa, and next as the panacea for ameliorating underdevelopment.

Of course, there are few activities more 'people-centred' than education, yet the curriculum content in Tanzania can be faulted for its ideological support of 'production-centred' modernisation ideals at the expense of, among other things, community based knowledge. Such an educational system is, by definition, representative of a knowledge system laden with cultural values and beliefs foreign to the societal context in which they are operating.

EXPOSING THE GAP: CONDUCTING RESEARCH FOR LEK

In attempting to explore the gap between local environmental knowledge and classroom-based knowledge in Tanzania, I utilised qualitative educational research techniques, and methods from Participatory Rural Appraisal or PRA. With PRA methods, researchers are encouraged to select from a 'menu' of techniques and adapt and apply them to different research questions or situations. In this paper, data gathered using three PRA techniques are presented.

THE RESEARCH FRAMEWORK AND APPROACH

Essentially I wanted to compare and contrast knowledge and ideas about the environment in 3 local communities (Morogoro, Sameh, Mtwara) with what I was able to observe and measure in 3 secondary schools. I focused on the environmental education components of the various curricula by observing classes and interviewing teachers and students at the Form V and VI level. Classroom observations helped provide an understanding of teaching styles and environmental education content at the present time. In the community, I focused on local people living immediately around the schools, interviewing and observing elder men and women primarily. PRA methods were used with both informant groups so that a more accurate comparison would be possible. Research with each respective group was conducted separately.

TECHNIQUES

Because I am presenting only a sample of the data collected in Tanzania, I am restricting the description of methodology to three techniques which, I would argue, not only provide data about LEK and school-based environmental education, but might also be useful pedagogical techniques for capturing and documenting LEK by students through a progressive environmental education curriculum. The description of each technique will be followed with presentation and discussion of the data.

Open-ended interviews

The majority of the data comes from lengthy qualitative interviews with local elders, as well as with small focus groups of students[6] and teachers. Figures 2 to 4 include a small excerpt from responses given by elders and students in open ended interviews. The responses from elders are listed in the left-hand column, and students in the right. Italicised statements indicate agreement between elder and students responses. Numbers appearing in brackets after a response indicates that more than one informant gave the same answer.

Q1. How long have you lived in this area?			
ELDER			**STUDENT**
Morogoro	**Sameh**	**Mtwara**	**Morogoro**
30 years	54 years	40 years	- average or less than 16 months for the group
21 years	78 years	73 years	**Sameh**
27 years	45 years	58 years	- average of 16 months
60 years	50 years	40 years	**Mtwara** - each of the six students gave an answer ranging from 5 - 14 years

Table 2. Elder and student responses in open-ended interviews

Discussion:

The different responses from student and elder informants is explained by the fact that, with the exception of Mtwara, students come from other parts of Tanzania and are boarded for their secondary school years. In Mtwara, students are local and responded accordingly. The fact that most students have relatively little experience in these communities seems to influence their responses

---

[6] It is important to clarify that students were asked to provide their answers on the basis of the 'knowledge' they had of the environment in the area of their secondary school rather than customs or beliefs that pertained to their home village.

significantly. Later, where students cite examples of local beliefs, and where these comments differ greatly from the responses of the elders, experience in the community becomes an important factor.

Q3. Can you tell me if there have been any changes in the local environment?

ELDER	STUDENTS
Morogoro - *since 1981, unreliable, insufficient rainfall* - *poor soil fertility* - *very low harvests* - charcoal dealers have cleared hills of forest - dried springs - rainy seasons are now very short - seasons are becoming very unreliable - change caused by decreased rainfall - occurrence of new diseases	**Morogoro** - *we were told that Mzumbe used to experience rains in Nov - Dec, but it is now February and it has rained lightly only twice.* - *soil fertility is declining due to bush-burning* - *there has been a decline in harvests and a rise in maize prices* - dirty water is causing severe disease such as typhoid, dysentery, and meningitis
Sameh - *wet seasons have all but disappeared.* - many years ago this was a high savannah with many trees, elephants, and plenty of Rhinoceros. - there used to be plenty of rain, two rainy seasons, and good harvests. - rainfall was good up to about 1969-70. Since then, soil fertility has declined, harvests are very low, and there are acute shortages of water affecting people and livestock(2). -greater incidence of human and animal disease but many of the traditional medicinal herbs have been destroyed. - forest cover has declined. Most trees have been cut down. High winds, and natural streams have dried up. - most of the trees in this area have been cut down	**Sameh** - *the students commented that the area had become continuously drier since they arrived.*
Mtwara - yes, changes in forest cover, disease, water supply, and *soil fertility(4)* -yes, especially decreased rainfall. Water supply is very scarce compared with the early 1900's. - soil is not fertile without the application of artificial fertilisers.	**Mtwara** - it seems to have become hotter - there has definitely been a decrease in annual rainfall - decrease in *soil fertility* - increase in population - outbreaks of disease

Table 3. Elder and student responses in open-ended interviews (continued)

Discussion:

In general, as my line of questioning focused more on the details of the local environment and environmental change, the responses of the elder informants became consistently more detailed and confidently delivered. During interviews, elders were more emphatic and dismayed when describing environmental degradation. Students tended to hesitate and confer with each other before stating that one aspect or another of the environment had indeed exhibited change. With respect to this question, the trend graphs developed between myself and the elder informants further corroborates the interview data. Environmental trend graphs are presented later in this paper.

Responses from Sameh are more divergent. Sameh was by far the most obviously denuded area of the three sites considered in this study, and the segregation of the school and community seemed most pronounced there as well. This is evident in the near total lack of comment from the student focus group in response to this question. Elders gave detailed accounts of the same bio-region historically and juxtaposed it with conditions today. They noted declining rains, disappearing streams, declining soil fertility, loss of traditional plants and traditional medicines, and continuing deforestation. Students, on the other hand, noted only that the area had become drier since their arrival at the school.

Comparisons of the elder and student responses from Mtwara reveal much more consensus regarding significant environmental change. It is important to note that Mtwara is the only school in the study where students are local, perhaps indicating that environmental concerns are shared by parents, students, and members of the community.

Q8. Tell me about customs or beliefs about the environment in this area?	
Morogoro - tambiko: sacrificing under the Mwongo trees, particularly at Mgetya and Bunduki lake where our forefathers used to go and offer sacrifice -even during the driest seasons, rain still seems to fall on Ukwagule forest, so it is considered a shrine of all the Waluguru tribe. - community members believe that they have no control of environment. Any change is taken as either the will of God, or their past ancestors. For this reason, people pray, sacifice their harvest, or slaughter an animal to please God or ancestors(2). - dwindling forests, drought, and disease are considered punishment for people's sins by ancestors and God. - most people, especially elders, believe that everything is under the control of God and when faced with environmental problems they pray more. **Sameh** - the beliefs concerning the forests and related rituals were prominent in this area. Now they are seen as archaic. Even plants with medicinal properties are not respected. - beliefs about forests have disappeared, though a couple of clans still have untouched forests. Close to Same, most trees have been cut regardless of 'clan forests'. Even big trees where women used to pray for fertility have been cut. **Mtwara** - natural forests are believed to be the source of medicine as well as rivers and streams(3). - the sources of rivers are believed to be protected by God. - because forests are the source of medicine, anyone cutting forests improperly would be punished by God. -forests are ancestral home where rainfall originates.	**Morogoro** *- no we can't because we are all newcomers to this area.* **Sameh** - the Mwarobaini tree grow in this area. These trees are good for about 40 different ailments. People plant them around their houses and tend not to cut them because of their medicinal benefits. **Mtwara** - big baobab trees used to be considered special, religious sites. - there were many traditional plants to protect people from evil spirits. Islam and Christianity now discourage these beliefs. - there is still much trading in medicinal plants

Table 4. Elder and student responses in open-ended interviews (continued)

Discussion:

There is evidence in Morogoro and Sameh that European or Christian belief systems have discouraged the passing down of beliefs about the forests or other natural phenomenon. Such beliefs were deemed fundamentally pagan but, nonetheless, are likely to have had some net conservation impact. As Gadgil *et al.* (1993) argue, indigenous systems of knowledge develop based on decision rules of which the social group as a whole may not be fully aware. For instance, certain individuals in a tribe or social grouping may be the primary informants of decision rules concerning the environment while other members of the group merely recognise the rule as a custom. Such a process "inevitably results in a commingling of knowledge, practice and belief which seems to characterise the conservation practices of indigenous people" (1993).

Again, the level of detail and consistent deference to environmental matters in the domain of 'God's will' was very clear in the responses from the elders. The commingling of knowledge and practice with belief needs to be first viewed objectively but not condemned by the outside observer, and then understood and perhaps decoded for the environmental ethic which pervades the belief. To do otherwise may lead to the characterisation of beliefs as mere superstition rather than as the codes of conduct for entire social groups, part of which concerns environmental conservation and resource use.

The use of certain trees as shrines to ensure rainfall, good harvests, and even fertility is a common theme in elder responses to this question. Taking this as one example, the custom of regarding these trees as religious shrines may have originated out of a perceived or prescribed need to maintain biodiversity (with respect to the tree species). On the other hand, the maintenance of biodiversity may just be a positive, yet unexpected, outcome of a purely spiritual dictate. In either case, the beneficial environmental impact of the belief in practice cannot be overlooked. Instead, the rationale behind the religious belief should be studied and understood.

Community time lines

Community elders and students were both asked to describe significant events in the community going back as far as they could remember. For some, this included droughts, epidemics, and political milestones. Every attempt was made not to bias responses in any way. Informants were free to highlight whatever came to mind, this way ensuring that the event was truly significant for them. The resultant timelines, particularly those of elder informants, indicate a detailed historical knowledge about key environmental events that would be interesting and instructive in a classroom setting. Figures 5 and 6 are timelines from an elder and another from a student focus group in Morogoro, for comparison:

DATE	EVENT
1939 - 45	- WWII
1940s	- Forced cultivation of tobacco
1973 - 74	- Great famine due to drought
1976	- A great traditional healer, Mkomma Mogella, died
1983 - 4	- Prolonged drought
1974	- Tazara railway passes Morogoro
1982	- Mindu dam was completed
1974 - present	- Agricultural activity takes place on Mongwe hills
1993 - 4	- Drought

Table 5. Timeline 12/02/94 - Morogoro elder, H.J. Jumbe

DATE	EVENT
1974	- famine
1992	- famine
1993	- very heavy blowing winds
present	- cattle dying, lack of grasses for feed

Table 6. Timeline 12/3/94 student focus group

Discussion:

It is clear that elders are able to provide more detailed historical timelines. The ability to discuss environmental change is the most intriguing aspect of this data. In order for students to understand processes of environmental degradation, it would be helpful to collect evidence of changes over time.

Environmental trend graphs

Both groups were asked to depict in graph form, the perceived trends of phenomena such as rainfall, harvests, food production, and forest cover over several decades. Invariably, the informants indicated a deteriorating situation. In comparing each respective informant group, I suspected that the student groups were extrapolating their graphs from their experience of the current situation - for example, the fact that the immediate area around the school was heavily deforested. The community elders on the other hand spent much time in the interviews describing the relatively healthy state of the natural environment some decades ago. Figures 7 to 9 were drawn with community elders in Morogoro, while figure 10 was done with the student focus group. The descriptive information alongside the graphs was provided by the informants.

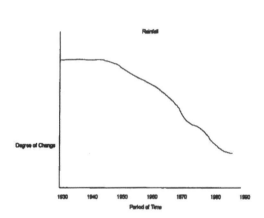

Comments:
In the 1960s, Morogoro gained municipal status. During the 1970s charcoal production was allowed on hill slopes. The 1980s saw increased population pressure on vegetation cover.

Figure 1. Environmental trend graph showing trends in rainfall

Comments:
During the 1960s - 70s, tobacco monoculture was encouraged. Mixed cropping of maize and legumes was introduced in the 1970 - 80 period. In 1983 - 86 chemical fertilisers were widely used, but from

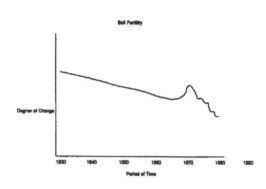

1987 - 90 there was reduced use of fertilisers due to dramatic price increases. From the 1990s to the present, there is a heavier reliance on farmyard manure.

Figure 2. Environmental trend graph showing trends in soil fertility

Comments:
From 1930-60, colonial measures restricted haphazard forest harvesting. During the post-independence period of the 1960-70s, the government issued permits to harvest forests unsustainably. Since 1980, illegal harvesting of forests for charcoal production has increased.

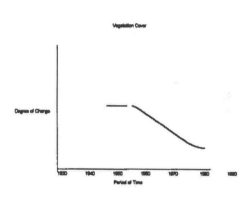

Figure 3. Environmental trend graph showing trends in crop harvests.
Source original fieldnotes, 10/01/94

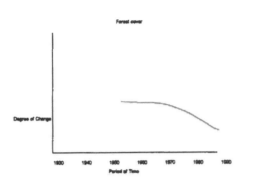

Comments:
No additional data given by
the students. Note that the
analysis spans a shorter
time period.

Figure 4. Environmental trend graph showing Mzumbe student data

Discussion:

These graphs provide another technique for expressing and comparing
knowledge of environmental change over time. Clearly, elders are able to
recount significant events over several decades, and link these events, in this
particular case, to consistent environmental degradation whereas students had
less to contribute. This research technique could be used by students doing
their own interviews with community elders.

SUMMARY AND CONCLUSIONS

The structure of education in Tanzania has persistently promoted the aims and
objectives of a modernisation paradigm of development which, among other
things, has resulted in the marginalization of local knowledge, particularly local
knowledge about the environment. The potential value of LEK in addressing
environmental sustainability is well established in the literature, and I have
hypothesised that LEK could be given some recognition in formal secondary
school structures through newly emerging environmental education curricula
materials. Research conducted in rural Tanzania over a 7 month period in
1993-4, indicates that the existing gap between LEK and school-based
curricula in 3 communities is significant.

There is a great deal of potential for environmental education to not only make a significant difference in ameliorating human impact on the planet, but to also respond in relevant ways to human needs for development. A type of environmental education that promotes sustainable living within increasingly fragile ecosystems is possible. In order for this objective to be realised, people must be speaking the same language. I do not mean Kiswahili, French, Shona, or English. The language I refer to is the language people use to articulate their respective understanding of their environment. This kind of knowledge transcends the boundaries of spoken vernacular and allows people to communicate and solve problems together.

The challenge is to first try and understand local or traditional knowledge in a way that permits its systematic translation to more formal settings, without losing the intuitive integrity inherent in vehicles such as story telling, songs, dance and metaphor. Programs must be designed and experimented with in terms of integrating classroom learning with community based sources of knowledge. With that experience, researchers and educators can re-investigate and re-apply LEK in classroom situations in a continuous cycle of research and application.

REFERENCES

Bakobi, B. (1992) 'The Dawn for Environmental Education, Public Awareness and Involvement', Paper presented during the workshop, *Incorporation of environmental education in school programmes 26-28 February*. Kibaha, Tanzania.

Berkes, F. (1993) 'Traditional Ecological Knowledge in perspective', In

Inglis J. T. (Ed.), *Traditional Ecological Knowledge:Concepts and Cases*. Canadian Museum of Nature and IDRC, Canada.

Buchert, L. (1992) 'The Community School as an Instrument of Social Innovation in Tanzania: An Analysis Based on Two Cases', In *CESO Education and Training in the Third World: The Local Dimension*. Vol.18, CESO, The Hague, pp. 123-143.

Fafunwa, A.B. and Aisiku, J.U. (1982) 'Education for Africa: Progress and Prospects', In Fafunwa, A.B. and Aisiku, J.U. (Eds), *Education in Africa: A Comparitive Survey*. George Allen & Unwin, London, pp. 254-261.

Fitzgerald (1990) 'Environmental Education in Ethiopia'. *International Journal of Educational Development*.

Gadgil, M., Berkes, F. and Folke, C. (1993) 'Indigenous Knowledge for Biodiversity Conservation', In *AMBIO*, Vol. 22, No. 2-3, pp. 151-156.

Gallopin, G.C., (1991) 'Human Dimensions of Global Change: linking the Global and the Local Processes', In *International Social Science Journal*. Vol. 130, November 1991.

Johnson, V. (1992) *'Environmental Education in Africa'*. Unpublished paper, Wildlife Conservation Society of Tanzania Dar es Salaam.

Lalonde, A. (1993) 'African Indigenous Knowledge and its Relevance to Sustainable Development' In Inglis J. T. (Ed.), *Traditional Ecological Knowledge:Concepts and Cases*, Canadian Museum of Nature and IDRC, Canada

Meadows, D. (1989) *'Harvesting One Hundredfold'*. UNEP, Kenya

NEMC, (1994) *'Draft National Conservation Strategy and Action Plan'*. I Internal Government document.

Nyerere, J. (1982) 'Education in Tanzania', In Fafunwa, A.B.,and Aisiku, J.U., (Eds), *Education in Africa: A Comparitive Survey*. George Allen & Unwin, London, pp. 235-254.

Nyerere, J. (1967) *'Education for Self-Reliance'*. National Printing Company, Dar es Salaam, Tanzania.

Osaki, K.M. and Wandi, D. M. (1994) *'Environmental education in the Formal and Non-Formal Education System in Tanzania'*. Unpublished paper prepared for an IDRC sponsored workshop on Environmental Education Research, IDRC, Nairobi.

Permpongsacharoen, W. (1992) 'Alternatives from the Thai Environment Movement', *In Nature and Resources* Vol. 28, No.2, pp. 4-14.

Redclift,M. (1987) *'Sustainable Development: Exploring the Contradictions'*. Methuen, New York.

UNEP (1976) *Connect* Vol. 1, No. 1. January 1976.

UNEP (1982) *Saving Our Planet; Challenges and Hopes*. UNEP, Kenya

Wolfe, J., Bechard, C., Cizek, P., Cole, D., (1991). *'Indigenous and Western Knowledge and Resource Management Systems'* unpublished, University of Guelph, School of Rural Planning and Development, USA

An earlier version of this paper was presented at an IDRC sponsored workshop on Research Issues in Environmental Education in Eastern and Southern Africa, held at the Silver Springs Hotel, Nairobi, Kenya, August 29 - September 2, 1994. Substantial additions were made to suit the purposes of this book.

Chapter 8

NETWORKING FOR ENVIRONMENTAL EDUCATION

Zena Murphy
European Research and Training Centre on Environmental Education
University of Bradford
Department of Environmental Science
West Yorkshire UK
BD7 1DP

INTRODUCTION

"Better communication and networking ... are required. Existing networks should be extended and new ones developed ... to encourage information exchange, share good practice, promote mutual understanding, and enable more effective strategic planning."

This quote comes from the 'Key Needs and Mechanisms for a Strategy' section of the Scottish Office document 'Learning for Life: A National Strategy for Environmental Education in Scotland' (1993). The increasing incidence of references to networking in such national documents / strategies / plans which go back as far as the Belgrade Workshop and even earlier, indicates the growing importance attached to such a practice. Here is an attempt to address some of the questions repeatedly asked about networking - what is networking, what does it mean, what types exist, how do computers come into it ('networking' being more traditionally associated with the world of computers), and what to do when things go wrong.

NETWORKS: DEFINITION AND FUNCTION

There have been many attempts to define the term 'network', including that by Goldstick (1992) who defines a network as "a system that links people together for the purpose of sharing information. Information is power, and a network spreads that power": Plucknett *et al.* state that "networks can be defined as associations of independent individuals or institutions with a shared goal" (in Sigma Xi, 1993), and finally Long (1993) sees a network as "a group of people, brought together by their willingness to work together on a common project, to achieve common goals and objectives".

Regardless of the definition that one prefers, the philosophy and essential processes of networking do not differ much between networks. Instead, it is the subject matter that differs. Therefore, the description of networking principles which follows can be applied to any and all of the current networks in existence. Networking essentially involves three main activities:

1. information exchange;
2. active participation in that information exchange, and
3. interaction between 'suppliers' and 'users' of information (Leal Filho, 1993).

As illustrated above, networking can and does go beyond the process of information exchange. Lane (1991) sees the concept of networking as a means of enabling the exchange of information "and for the co-ordination of services and activities".

The basic process involved in networking is the continuous two-way flow of information, between people who can, at different times, and using different methods, act as both 'suppliers' and 'users' of that information. 'Information' in this case can mean anything from news about the latest environmental 'goings-on' (for example the latest problem or crisis nationally, internationally or locally), or the latest ideas or philosophies concerning 'best practice' with regard to solving or overcoming a particular problem: it all depends on the network's subject matter and essential concerns.

Different authors have offered different sets of goals / objectives / ingredients to ensure that a network is successful. Goldstick (1992) when offering some criteria for a successful network says that a network should:

- clearly define what is relevant;
- network only, not carry out other functions such as research and training;
- maintain a low profile so that other organisations are promoted;
- choose effective ways of communicating, and
- not simply receive and send but actively seek out relevant information.

On a critical note, Goldstick's criteria may perhaps be a little too limited to accurately reflect today's increasingly sophisticated networks, some of which offer an impressive range of activities which are lead by their members' needs and wishes.

DEFINING FEATURES OF NETWORKING PHILOSOPHY

Barton (1994) describes what he sees as the five major principles which are common to the UK's Regional Environmental Networks. This is a useful list, and is featured here to illustrate the defining features which synthesise the networking 'philosophy':
1. ownership;
2. consensus;
3. neutrality;
4. communication;
5. mutual value.

1. *Ownership*: networks should have no blueprint for ownership, as members are and should be independent, retaining their separate identities (and hence separate goals and objectives as members of the network) under the network umbrella.

2. *Consensus*: all networks are established by consensus, as a result of a perceived need for the network. There is no hidden agenda for the network: it exists simply to be a catalytic form of information exchange, operating by mediating between interest groups.

3. *Neutrality*: no attempt should be made to coerce members into action: participation is offered equally to all members, and active outreach is practised to involve new and non-members. A network can also act as a forum for conflict resolution, and aims to develop a vision which can be shared by all members.

4. *Communication*: first and foremost, networks exist to act as a communication channel, providing information through such means as a newsletter / a central office / meetings and events. A network also promotes good practice of information exchange, and acts to market itself as an exchange location for information, resources etc.

5. *Mutual value*: all networks require and promote mutual respect among members, aiming to improve understanding between interests where the environment is concerned. Networks are and should be non-competitive, providing support and reinforcement to their members, aiming to be locally responsive and promoting inclusiveness in their practices and participation.

NETWORKING FOR ENVIRONMENTAL EDUCATION

With specific reference to why networking is being so widely advocated and used as a tool for environmental education, Leal Filho (1993) is in no doubt that the establishment of networks can help the development of co-operation in international environmental education, particularly in the following ways:

a) by putting people in touch with each other, catalysing discussion and the establishment of links;

b) by disseminating information on current facts and on matters of common interest such as training opportunities, and the undertaking of joint projects and events, for example workshops and training sessions;

c) by fostering the use of successful techniques and ideas in other countries, thus contributing to the development of environmental education in nations where it is not well developed.

These ways can all be described as tangible benefits in the development of environmental education, assisting with its growth world-wide in a cohesive, progressive manner. But there are also other, less tangible benefits which arise out of networking. Basu (1991), describes these benefits "... it leads to the development of a community of institutions by ending their isolation. Solidarity, mutual support and a united front which is gained leads to an effective orchestrated action". Chumo (1992) points out that networking also discourages the duplication of environmental education efforts (which have been identified in the past as a problem in environmental education efforts world-wide - see Basu, 1991), and, due to the ever-changing nature of environmental issues, which requires educators and learners to continuously expand their scientific, ethical and technical knowledge to cope with these changes, networks promote training through the sharing of experiences and case studies.

TYPES OF NETWORK THAT EXIST

There are a number of different types of networks. These different types can be classified according to:

1. the level on which the network operates;

2. the region in which the network operates, and
3. the information exchange medium the network uses.

Types of network 1: Levels of networking

According to Leal Filho (1993) networking can be described as taking place at three levels:

- individual - individual: between people who are active in this field and who wish to further their knowledge and range of contacts by liaising with people sharing the same interest;
- individual - institutional: between an individual(s) and an institution(s) through which both parties share their knowledge towards a common goal;
- inter - institutional: between institutions concerned with an issue and active in the undertaking of projects. In such circumstances the expertise available at one institution on a particular topic may be of interest to the other and *vice versa*.

In the context of the above, some variations may be found, but whichever the case, the likelihood of success of any co-operative link will depend on the reliability of the communication links established between the concerned parties or individuals, the frequency with which the exchange of information takes place, the care in reaching the pre-set goals and the awareness, from each party, of his / her individual contribution to the process.

Types of network 2: Geographical remits of Networks

There are 4 broad divisions when one considers the geographical remit of networks:

1. *International networks* tie together institutions or individuals from different parts of the world with a common purpose to share information and resources (Basu, 1991). For example, the 'Global Rivers Environmental Education Network' (GREEN) links countries together for the purpose of monitoring and conserving rivers and improving or introducing environmental problem-solving skills.

2. *Regional networks* are concerned about topics and problems affecting a particular geographical area, for example the 'ASEAN Region Network for Environmental Education' (ARNEE) has a geographical remit of the ASEAN countries: Brunei, Malaysia, Singapore, Thailand, Indonesia and the Philippines (Vietnam is to join shortly).

3. *National networks* work towards a national goal, for example the 'UK Forests network', as its name suggests, works primarily on UK forestry-related issues.

4. *Local networks* work within a narrow geographical area, for example 'Links for Leeds', which works within a small area in West Yorkshire, UK. Such networks have a much smaller remit, encouraging close working relationships between individuals and institutions in a limited area.

Types of network 3: information exchange media

Networks can be said to be in one of three classes when it concerns which information medium is being used:

1. either the network uses the printed media to exchange information, for example through the distribution of:
- newsletters;
- bulletins;
- magazines;
- periodicals; or

2. it makes use solely of an electronic medium, including
- telephone;
- fax;
- cable;
- telex;
- electronic mail; or

3. a combination of both types to a lesser or greater degree.

As research undertaken shows, this latter type is predominant (Murphy, 1994), as each network makes full use of the appropriate and most convenient media. Many realise it is not good practice to rely on one type of medium only: the exclusive use of, for example, newsletters can sometimes miss an audience, or have a substantial delay in reaching its destination. The most appropriate method(s) of communication is network-specific, which further depends on the needs and ability to access information, of the network members.

Information exchange within a network ideally works as a two-way process: information flows from the network to the members and members also feed information back to the network: the information exchange structure within a network determines how much of a two-way process (or feedback loop) this really is.

If exchange is through a co-ordinating body, such as a central information exchange service, (the information being sent initially to the central body from the members), this can be illustrated diagrammatically by a wheel with spokes radiating from a central hub (see Figure 1). The hub represents the central information exchange service or person and the spokes and outer rim the network members.

One drawback to this form of structure however is that there is often a delay between the central information exchange body initially receiving information and the members receiving it, which may mean the information is out-of-date by the time they receive it.

As mentioned earlier however, networks essentially exist to promote the free flow of information and any withholding of information from and between members can be seen as the antithesis of the networking philosophy. A contrasting information exchange structure, as illustrated by the pilot Hampshire Environmental Network (pHEN) (Hampshire County Council 1994), and also by the longer-established Regional Environmental Education Forums (REEFs) in Scotland (SEEC, 1993) is that of an open and growing web with no clear centre (see Figure 2). Each strand is directly connected to every other strand, all information therefore being accessible to all.

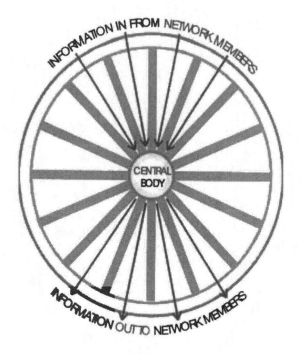

Figure 1. Information exchange through a central body

NETWORKING USING ELECTRONIC COMMUNICATION

Rohwedder (1994) says that "computer-aided environmental education may prove to be our most powerful educational tool for promoting everything from environmental awareness to environmental action". Using computers implies two possibilities: the use of computers as a medium on which to work (i.e. word processing, role paying, database use), and the use of computers to obtain information from a variety of sources, in other words, electronic mail or (e-mail).

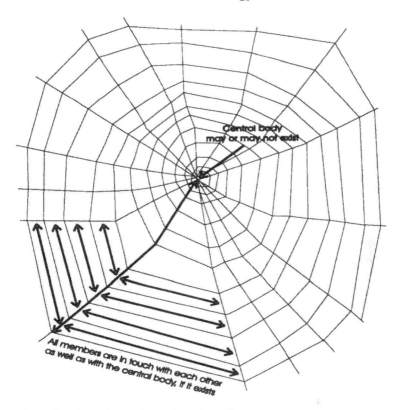

Figure 2. Information exchange through a web

As pointed out by Ngola (1995), electronic communication is within the reach of any individual or office that has a computer, a phone line and a modem. The modem translates the text (digital signals) into electrical signals that can be sent through a phone line. At the other end, a 'node' computer translates the signals back from electrical to digital signals via a modem.

While it may seem to be a complicated process, e-mail communication takes very little time at all, and is also reliable (dependent upon whether systems are functioning). It is the cost however that is the most attractive advantage of electronic communication. Once the initial costs of software and training are met, and providing the user already has a computer, an e-mail

143

message costs as little as one tenth that of a conventional fax. This makes electronic communication a excellent way to save money, especially for NGOs with large international communication expenses.

Also available, reports George (1995) are the numerous networking sites run by companies: the Internet (only one example of an increasing number of such sites) was developed in the US and is a group of 'hosts' which can be used, for example, to send e-mail, download software, or access discussion lists or bulletin boards on a rapidly increasing and bewildering variety of topics.

However, George finds, navigating through the 'Net' and getting what you want can be "as confusing and frustrating as being an Englishman trying to eat rice with chop-sticks." Addresses for individuals are not easily obtainable, and often out-of-date; a dot in the wrong place is 'fatal'; there is no guarantee, should your mail not reach its destination, that you will be informed. The bottom line, George informs us, is that it is not 100% reliable, and that hardware necessary for the job is expensive.

Lane (1991) emphasises that "the concept of environmental education networking is proposed as a means of enabling communication between individuals and groups: for the exchange of information and for the co-ordination of services and activities. Computers are not necessary to support this intent but there is no doubt that they can make it easier".

Electronic communications are becoming increasingly sophisticated and more easily accessible, but it would be unwise, George warns, to rush out and buy the necessary 'surfing' equipment without appreciating the problems and perhaps waiting for the hype - and more importantly the prices - to subside.

EXAMPLES OF ENVIRONMENTAL EDUCATION NETWORKS AT DIFFERENT GEOGRAPHICAL LEVELS

The following examples have been chosen (from a list of many) to provide a brief overview of the main audiences, activities, objectives and information exchange methods of networks at different geographical levels.

On a local scale, the 'Manchester Environmental Education Network' is a new networking initiative, established only in 1994, based in schools in the Manchester area. The network as described by Levy (1994) arose out of the Global Forum event which took place in Manchester in June of that year. The basic ethos of the network is envisaged as being one of ownership (of the network's success), promoting sharing, aiming to be friendly and fun, inspiring, stimulating, growing, complementary and supportive of all environmental educators (teachers / key workers / outdoor education officers etc.), in Manchester.

In terms of communication, a newsletter named 'Beehive', with ideas from the steering group's and members' own experiences, is being produced termly and is distributed locally.

The network was established by first setting up a Steering Group, over half of which is made up of local teachers (from all teaching levels), plus a range of local organisations with different interests. Contained within this Steering Group, is a Working Group to work on specific policies, materials, activities etc., and to play an active role in sustaining and co-ordinating the network.

An Environmental Education jamboree was held in October 1994, inviting environmental education and environmental action organisations to provide information on what they do and can offer to schools, and since that time 'twilight workshops' (held in the evenings or weekends for teachers with an otherwise busy schedule), have been run by members or other organisations on a voluntary basis at little or no cost.

Future events are planned, all with an emphasis on preparing teachers for improving their skills to undertake active environmental education. Such training workshops aim to reflect the needs of the members. Also planned is the linking up with other such networks, for example the 'London Environmental Education Forum', and making the network sustainable.

Working on a national remit is the 'Environmental Education Network of the Philippines' (EENP) (described by Basu, 1991), which has been under development with assistance in the form of an initial grant from the Ford Foundation. This grant assures the continued operation of the network during the organisation phase.

The network brings key regional colleges and universities in the Philippines into contact with colleagues and programmes of environmental education and awareness both locally and abroad. The goal is to establish long-term national linkages among educational institutions, between educators and NGOs, between environmental institutions, non-governmental institutions, the government's Department of Environment and Natural Resources (DENR) and other agencies interested in environmental problems. The objectives of the EENP are:

1. to create a network of regional colleges and universities, research centres, and NGOs for co-operatively promoting the advancement of environmental education and awareness in the Philippines;

2. to develop collaborative activities that will lead towards a sustainable management of the country's natural resources, and
3. to co-ordinate instruction, research, and extension initiatives among environmental and educational institutions in the country and to provide mechanisms for linking up these initiatives with global and regional (Southeast Asia) environmental programmes.

The network is a loose organisation of environmental institutions, with a permanent secretariat based at the Institute of Environmental Science and Management, University of the Philippines. Communication between members is in the form of a newsletter, and meetings and events are also organised, through which informal communication is possible.

The activities of the network are initiated in a flexible manner, to accommodate changing needs as new problems and priorities are introduced. The long-term plans of EENP are to consolidate all co-operative efforts by individual colleges and universities, research centres, and NGOs in the development of environmental education and awareness programmes for the Philippines.

Other examples of national networks are the 'Réseau Ecole et Nature', the French environmental education network, and 'EECom, the Environmental Education and Communication Network', based in Canada.

Regional networks include the 'Regional Environmental Education Forums' (REEFs), based in and working throughout the Scottish regions, the 'Environmental Education Network' (EE-NET), Latvia, which works in eastern European countries, the 'African Association for Literacy and Adult Education' (AALAE), working from Kenya, 'SHARE-Net' (South Africa), and the earlier mentioned 'ASEAN Region Network for Environmental Education' (ARNEE), based in Indonesia.

A very well-known example is the sub-regional 'South and South East Asia Network for Environmental Education' (SASEANEE), based in India. It was set in motion jointly by the IUCN Commission on Education and Communication and the Centre for Environment Education (CEE). A three day workshop was held in 1993, aimed to exchange ideas, information and materials on environmental education (CEE, 1993), and to discuss the operations and structure of the network.

A currently twice-yearly SASEANEE Circular reports on efforts in the region and outside, and reminds network members of their identity. Its main activities are the Environmental Education Banks, which have been set up in Ahmadabad and Bangolore with the aim of helping in the development of situation-specific teaching material. Supported by the International

Development Research Centre (IDRC), it is a collection of environmental education programmes and materials from all over the world, and can be accessed through workshops organised for small user groups such as teachers. The SASEANEE Directory (currently under development) is an add-on directory containing addresses and descriptions of individuals and agencies involved in environmental education in the South and South East Asian region.

Other periodic meetings, training events and regional network member internships (at CEE) are proposed, and the various agency members proposed the creation of model programmes and prototype materials (for example practical guidebooks). The CEE already has many such activities within its framework, which will be made available to the network, as it has nodal agencies or organisations in the network regions.

The 'Global Rivers Environmental Education Network' (GREEN), USA, an example of a network working at an international level, is an initiative of the University of Michigan (USA), whose objective is to promote environmental education by means of an hands-on approach: in this instance the monitoring of rivers and watersheds.

The goals upon which GREEN is based include helping learners to develop skills needed in environmental problem solving; working in groups; gathering, analysing, synthesising and interpreting information; clarifying norms and values; designing, implementing and evaluating a plan of action; and joint critical decision making. (Sneddon, 1992). In the environmental education goals of GREEN, some type of action-taking or problem-solving ideally follows the students' investigations and discussions of their watershed.

The GREEN Newsletter is the main instrument of communication between GREEN's office in Michigan and the participants, which are school teachers, university staff and staff of field centres in dozens of countries. The network study action group developed a survey for environmental educators which was posted on EcoNet, a global computer network with conferences for the sharing of information.

A study group focuses its investigations around ways of encouraging teachers and learners to include action-taking as part of their water quality monitoring program (Sneddon, 1992). A number of projects, as reported by Cromwell (1992), include the Partner Watersheds program which involves 14 countries. Through this program, the GREEN office matches classrooms of different cultures to carry out a GREEN project together, using the Partner Watersheds Manual that GREEN has created.

Computer Conferences, active throughout the Association for Progressive Communications (APC) are starting to become more active, with participants on-line from Germany, Russia, Australia, Japan Canada and the

US. They will continue to research improvements for the system, looking at ways to simplify use, and also potentially create an easily workable database of water quality data for students to use.

GREEN is also working to create a Manual for Low-Tech Water Monitoring. The purpose of this text will be to provide instructions on how to create a monitoring kit to enable groups to conduct tests for a general evaluation of local water quality.

GREEN Outreach is in the form of staff from GREEN offering workshops in Nova Scotia, Canada, Portugal, Australia, Montana, Texas, Illinois, Florida and Washington State. In addition, a national GREEN programme is being developed for Italy. Workshops in using telecommunications in the classroom were also presented in Ann Arbor.

Other examples of international / global environmental education networks include the 'North American Association for Environmental Education' (NAAEE), the 'International Network in Environmental Education' (based in Greece), the 'Alliance for Environmental Education' (AEE), and 'Globe NEEC - the Global Network of Environmental Education Centres', both based in the USA.

PROBLEMS OF NETWORKING AND POSSIBLE MEANS OF OVERCOMING THEM

Basu (1991) and others (Leal Filho, 1993; Hampshire County Council, 1994; Palmer, 1993) identify many obstacles to effective environmental education networking. A range of potential problems are identified here.

Firstly there is the culture of co-operation. Overcoming the (often unconscious) reluctance to share information can be a difficult process, and it may take time to establish trust between network members. Institutions with a strong tradition of working in isolation may attach undue importance to their individual existence in the network, and each entity may believe in the uniqueness of its own problems and needs. This problem is experienced, according to SEEC (1993) by the REEF network in Scotland.

The possible means to overcome this is, when establishing a network, to formulate a clear statement of the network's aims and objectives, to allow a baseline of understanding of the network's goals to grow among members, which can eliminate this possible barrier of a mismatch between members aims, their organisations aims and the network.

Secondly, there may be inadequate resources for effective networking. Constant efforts have to be made towards fund-raising, not only to ensure the long-term sustainability of the network (for newsletters, training events, salaries and so on), but also to keep network running costs (for example telephone, fax, and printing) down.

Taylor (1994) identifies four possible ways of generating income for a network, either to cover running costs or to allow expansion, many of which have been used by the Share-Net environmental education network, based in South Africa:

1. developing and selling to members low cost resource materials;
2. consultation work;
3. selling courses or facilities to members;
4. administering funds for projects - the network can either charge a service fee or benefit from the funds' interest.

Because of the wide geographical area covered and the possible limitations in a country's communications system, there may be problems in ensuring fast, effective communication exchange and in the reliable dissemination of information. Also, reliance on a single communication medium can lead to problems of information 'missing' many potential audiences of the network. Another potential problem is the break-down of a single communication vehicle, which could mean an interruption in information exchange, which ultimately lets the members down, and compromises a network's principles. To overcome this problem, using a wide variety of different locale-specific forms of information exchange might reduce the danger of interruption to the exchange process.

Finally, networks tend to be fragile. Many emerge out of personal contacts, and are spontaneous in their establishment, which mean they can collapse easily. Gayford (1992) suggests that it is important not only for networks to develop to meet the genuine needs of the providers of environmental education, but that mechanisms for the effective co-ordination of these networks should also be established and these too should be widely understood by users at different levels.

These and many other possible problems ideally need to be visualised in advance in order to take effective steps to overcome the obstacles to effective environmental education networking.

CONCLUSION

In conclusion, networks can be seen as useful tools to assist the environmental education process: useful in the basic fact that they put in touch environmental educators from all over the globe who can benefit from the ideas and experiences of others. However, networks should not be seen as the end in itself - they exist as one part of the process and, if sensitive to information and their members' needs, can provide a useful international input to local environmental education initiatives and *vice versa*. In the future, it is important that there should be even more international co-operation: networking will be an essential part of that co-operation, for benefits which will be felt at all levels.

REFERENCES

Barton, P. (1994) 'Principles of Regional Environmental Networks', presented at the *Environmental Networks Conference 1994*. IBM South Bank, London.

Basu, C.K. (1991) 'Developing Environmental Education Networks' In UNESCO *International Consultation Meeting on the Development of Networks in Environmental Education 4 - 8 February 1991*. Final Report, organised by the Colombo Plan Staff College for Technician Education, in collaboration with IEEP, Philippines, pp. 45 - 59.

CEE (1993) *A Workshop towards a South and South East Asia Network for Environmental Education, Recommendations*. Centre for Environment Education, India.

Chumo, N. (1992) 'Potential for environmental education networking in the region' In Lindhe, V., Goldstick, M. *et al Environmental Education - experiences and suggestions, report from a regional workshop in Nyeri, Kenya, 4 - 10 October 1992*. SIDA Nairobi, pp. 43 - 44.

Cromwell, M. (1992) 'A GREEN report from Ann Arbor' In *GREEN update, the Global Rivers Environmental Education Network*. Vol. IV, No. 3, 1992, University of Michigan, USA.

Gayford, C. (1992) 'Environmental Education Networks in the UK' In, UNESCO *International Consultation Meeting on the Development of Networks in Environmental Education 4 - 8 February 1991*. Final Report, organised by the Colombo Plan Staff College for Technician Education, in collaboration with IEEP, Philippines, pp. 70 - 74.

George, P. (1995) 'The Internet - the ultimate in networking?' *In Communicating Conservation*. Issue 3, May 1995, ICCE, UK, pp. 13 - 14.

Goldstick M. (1992) 'Some International Environmental Networks' In Lindhe, V., Goldstick, M. *et al Environmental Education - experiences and suggestions, report from a regional workshop in Nyeri, Kenya, 4 - 10 October 1992*. SIDA, Nairobi, pp. 41 - 43.

Hampshire County Council (1994) *Working Together in Sustainable Networks (summary report)*. Hampshire County Council, UK.

Lane, T. (1991) 'Computers and Environmental Education Networking: Use, Functions & Implementation Strategies' In UNESCO *International Consultation Meeting on the Development of Networks in Environmental Education 4 - 8 February 1991*. Final Report, organised by the Colombo Plan Staff College for Technician Education, in collaboration with IEEP, Philippines, pp. 60 - 64.

Leal Filho, W. (1993) 'The role of co-operation and networking in the development of international environmental education' In University of Indonesia *Proceedings of ASEAN Region Conference on Environmental Education for Sustainable Development, June 2 - 5, 1993*, Jakarta, Indonesia, pp. 2 - 9.

Levy, R. (1994) 'The Manchester Environmental Education Network', In Murphy, Z. *Networks for Environmental Education: Proceedings of a National Seminar*. University of Bradford, UK, pp. 16 - 19.

Long A. (1993) 'How to network: a beginner's guide' In *Med Campus Echo Issue 3 October 1993*. Med-Campus Technical Office, Rome.

Murphy, Z. (1994) *'Environmental education and networks in the Commonwealth'*. Unpublished MPhil thesis, University of Bradford, UK.

Ngola, M. (1995) 'Electronic Communication' In *Communicating Conservation*. Issue 3, may 1995, ICCE, UK, p. 8.

Palmer, J. (1993) 'Goals and conflicts in university based environmental education centres' In Mrazek, R. (Ed.) *Alternative Paradigms in Environmental Education Research*. North American Association for Environmental Education, Ohio, pp. 205 - 226.

Rohwedder, W.J. (1994) 'Using computers in environmental education' In ICAE *Adult Environmental Education: a workbook to move from words to action*. ICAE, Canada, p. 40.

SEEC (1993) *Networking Environmental Education: Regional Environmental Education Forums in Scotland*. Scottish Environmental Education Council, Stirling, Scotland.

Scottish Office (1993) *Learning for Life: a national strategy for environmental education in Scotland.* Scottish Office Environment Department, Edinburgh.

Sigma Xi (1993) *International Networks for addressing Issues of Global Change, summary report of a workshop.* Avila Retreat Centre, Durham, North Carolina, August 24 - 25, Sigma Xi, USA.

Sneddon, C. (1992) *GREEN, The Global Rivers Environmental Education Network.* Volume IV, No. 2 1992, University of Michigan, USA.

Taylor, J. (1994) 'A Case Study of Share-net', report compiled for *Workshop on Environmental Education for Youth 14 - 19 March 1994,* SADC, Windhoek, Namibia.

Chapter 9

ENVIRONMENTAL EDUCATION:
A PROJECT APPROACH

Chris Taylor
European Research and Training Centre on Environmental Education
University of Bradford
Department of Environmental Science
West Yorkshire UK
BD7 1DP

INTRODUCTION

Environmental education projects are extremely important instruments for environmental education as they take environmental information into communities. One may observe an increase in the use of projects as an approach to delivering environmental education since the 1970s onwards. Environmental education projects are characterised by their specificity, in terms of beneficiary groups, aims and objectives, also being diverse in nature. The variety of educational approaches used in projects is similarly diverse. The diversity of approaches allows information to be delivered in a way which is commensurate with the beneficiary group. In this context, informal education methods can reach both non-literate and literate group members. Access to environmental education is therefore increased, where normally it has been limited to the formal education system.

Projects are distinct from environmental education programmes in that the scope of the initiative is limited in terms of objectives and duration. It is not uncommon for projects to be incorporated as initiatives within a programme.

The potential for environmental education projects is enormous. The number of environmental education projects currently on-going appears to be limited by the scarcity of funding and resources. Monitoring and evaluating environmental education projects is an essential tool to not only inform management of the progress of a project, but also to convince donors of the merits of such initiatives.

This chapter will examine the common features of environmental education projects, address the potential of projects in the context of developing countries and provide some examples of this approach to delivering environmental education.

RATIONALE BEHIND ENVIRONMENTAL EDUCATION PROJECTS

Whilst environmental education in the formal education system is an essential source of environmental knowledge, it is evident that the proportion of the population of a developing country having access formal education is generally limited to children. It is recognised that everyone should have access to environmental education, and it is in this context that environmental education projects play a complimentary role to formal environmental education.

A salient feature of environmental education projects is the extent to which they are designed to deal with a specific environmental topic, or they are designed to address specific local environmental issues within a defined area. The ability of environmental education projects to meet the immediate needs of an identified group is a fundamental aspect. It is apparent that environmental education projects can be used as a problem-solving tool in a specific area, as well as one dealing with environmental issues on a wider (possibly national) scale. The flexibility allowed for in an environmental education project, and the variety of methodologies used, enables an environmental education project to convey messages in the most appropriate format.

In many developing countries, the role of environmental education projects in bringing environmental education to the whole population can equal any other approach to environmental education.

THE BENEFITS OF A PROJECT APPROACH

Projects have existed for many years, although it is only recently that they have been given the title of environmental education projects. Most development projects involve some degree of environmental education, and in this context, may be regarded as an additional source of environmental education projects. It is common that any type of education (especially in developing countries) that addresses the way in which people live, and how they may 'improve their lives' inevitably involves education concerning the way in which people utilise the environment as a resource.

It is apparent that environmental education projects as a specific type of project have increased in number relatively recently, and it remains common for many projects to address issues of development and poverty in the process of imparting environmental education, and changes in behaviour.

A salient feature of environmental education projects is that they are specially designed with a given beneficiary group in mind (in a specific region). The identification of all groups involved in a specific area requires a considerable level of thought. This is essential, as the result is a project which is highly relevant to the beneficiary groups, and should therefore act to motivate them. The design of a project should be achieved in conjunction with the various beneficiary groups. Environmental education projects usually address the immediate needs of the beneficiary groups within the confines of the local environment. As a result of this, readily tangible outcomes may emerge from projects, thus providing incentives to participate.

Not only should the participatory approach engender a feeling of empowerment among the beneficiary group, it should also motivate the group to become involved in the project, and hopefully, to use the idea of the project as a vehicle to bring about change long after the project has ended. The skills of organisation, administration, holding public meetings, and making decisions democratically are often indirect, but important benefits of environmental education projects.

Schools are commonly the only source of formal education although children learn from their cultural environment, their. Environmental education projects can be developed to provide education for a variety of age groups, from children to those who have left school (possibly many years ago). In a number of countries, the adoption of Western structures of formal education has resulted in what has become commonly termed the 'diploma disease'. The desire for formal qualifications occurs at the expense of 'learning for life' skills (that could be more immediately valuable). Generally, education concerning how to live in accordance with the local environment is included in the latter, non-qualification based education. As a result, much environmental education is lost in schools. The school cannot be relied upon to be the sole provider of environmental information / education and therefore environmental education must be introduced via alternative routes.

Bearing in mind the relatively low school-leaving age, and the high rate of school drop-out in developing countries, there appears to be large expanse to be filled in terms of environmental education. A classroom environment is not generally conducive to effective learning in all but the highly educated, and therefore projects must adopt alternative strategies conducive to effective learning. This should also take into account the needs of the illiterate and semi-literate where appropriate. Here lies the *forte* of environmental education projects. They can utilise non-formal and informal methods of education commensurate with the beneficiary group in order to relay environmental education / information. There is a whole multitude of techniques that can be used, (e.g. media including television, radio - especially effective in poorer developing countries - group meetings, poster sessions, talks, seminars etc.). Evidence of techniques used in specific case studies can be observed in the appropriate section.

THE LIMITATIONS OF ENVIRONMENTAL EDUCATION PROJECTS

Whilst the specificity of environmental education projects can observed to be a positive feature, that feature alone can limit the application of the project in other areas. A project on issues of water use in rural areas, for instance cannot be directly transcribed and used for a project on the same subject in an urban area. It is evident that the approaches found suitable in rural areas may not meet with a similar degree of success in urban areas. This may be especially true where one is relying on family / community responsibility to motivate a change in behaviour; in rural areas these types of approach may be successful due to the societal structure of the community, whereas in urban areas, the differing societal structure (e.g. transient, young, single workers) can render such approaches ineffective.

It is not uncommon for those groups of persons at the lower end of a community's social scale to be less responsive to project initiatives than those groups who are at the opposite end of the social scale (for reasons including education, employment and social structure). Whilst the latter groups may appear more enthusiastic, it is essential that extra effort be directed towards the groups of persons who are responding less favourably to the project.

Within the above context, the issue of motivation can possibly be the greatest hurdle facing the implementors of an environmental education project. Without generating the appropriate motivation, the project is likely to fail. It is at this point that factors such as identifying the various groups, communication methods, dealing with the immediate problems of the groups etc., need to examined in great detail. A failure to successfully motivate all groups involved in a project can severely restrict the chances of sustainable project success.

COLLABORATION WITH APPROPRIATE ORGANISATIONS

In many communities there is likely to be a number of organisations, e.g. women's groups, farmer's co-operatives, the church etc., that could be possibly considered for collaboration. In any given community, there may be one organisation that has become a familiar and popular organisation, and is one in which the local community have faith e.g. the church. The organiser of an environmental education project may consider approaching such organisations to facilitate and compliment the work of the implementing agency. There are a number of examples of the power that the church can have over communities.

Approaching an existing organisation does not reduce the work involved in setting up the project, but it may lead to a rationalisation of resources in terms of appointing local personnel to carry out non-technical functions. Because the existing organisation is trusted by the community, the introduction of ideas concerning a change in behaviour may not be judged with the same degree of scepticism that it might be if it has been introduced by the unfamiliar implementing agency. Working with existing organisations exhibits a willingness to consult other organisations, rather than possibly working against them. A hostile local organisation can have negative effects on the success of a project.

PROJECT STAFFING AND TRAINING

The majority of staff structures of environmental education projects tend to be characterised by a few, extremely dedicated staff undertaking all duties. The main reason for this is related to funding. The funding required for the appointment and training of extra staff is considerable. The ultimate consequence of this is a reduction in the quality of work of the project.

Specialist staff may be used where necessary: an example of this may be observed in project evaluation, or when dealing with topics with which the existing project staff are not familiar.

In some instances, it is appropriate to select and train local personnel to carry out some of the extension work of the project. The selection of the appropriate personnel is fundamental to the effectiveness of the appointment. An ill-selected local person may fail to fulfill his or her role, or they may use their title to disseminate additional information which is not related to the aims of the project. The same principles apply when approaching suitable organisations with a view to collaboration.

Having the expertise and resources to train additional staff members may be another problem. Where given appointments were not included in the project proposal, then finding the financial resources to appoint other staff members is extremely difficult.

BARRIERS TO IMPLEMENTATION

Initiatives for environmental education projects come from a wide spectrum of ogranisations. It appears that all but the most powerful non-governmental organisations (NGOs) are able to consistently raise funds to undertake environmental education projects to the extent to which they would wish. Smaller NGOs do not appear to have the same ability to convince donors of their worth (author's opinion). The reasons for this may be manifold. A lack of awareness of potential donors, lack of information as to the closing dates for project proposals, lack of familiarity with writing proposals (including appropriate budgeting) and lack of familiarity with organising environmental education projects may be just some of the reasons for the lack of small NGOs successfully securing funds for environmental education projects.

The larger NGOs (national and international) and government ministries seem to have a greater level of success in securing funding for environmental education projects. Whether this is for the converse of the above suggested reasons is not clear. What is clear, is that the funding for environmental education projects is extremely hard to secure, and takes a great deal of effort on behalf of the implementing agency to achieve. It is taken for granted that good relations with all potential donors will augment the chances of success. Keeping potential donors updated with details of current activities, and evidence of support from other reputable donors is all a part of maintaining good relations, and frequent contact.

A potential donor agency is likely to have a number of proposals for environmental education projects, and only a limited budget which has been allocated to the 'environment'. It is evident that donors want to fund only those projects which appear to be good idea, to be necessary, and a proposal which has been concisely written, with a budget which has been equally as well prepared. The budget of any proposal is the crux of success or failure (once the project has been accepted through the preliminary stages). The need for preparing good proposals is therefore essential if an environmental education project is to succeed in being funded.

The specific nature of environmental education projects means that the consumption of resources in terms of personnel, materials, transport etc. are considerable. It is not always likely that an implementing agency will have access to all the required resources. In remote areas, all of the above can be considered to be problematical, but in general terms, the problem of personnel remains almost universal (in all but the most well-funded of organisations).

Having the means to produce materials in a region without incurring excessive transport costs is just one example of the logistical problems surrounding the geographical location of an environmental education project. Where such provisions are distant, one must look towards alternative methods of disseminating information, i.e. ones which require the minimum of material production.

THE NEED FOR THE MONITORING AND EVALUATION OF ENVIRONMENTAL EDUCATION PROJECTS

At any given time, in any country, there is likely to be a number of on-going environmental education projects. If one adds to this the list of past environmental education projects, then one may consider that there is a wealth of knowledge and experience in running environmental education projects in a country. To what extent these experiences are shared is questionable. Access to literature regarding past and present environmental education projects is limited (in most instances) - especially where the environmental education projects have been located in remote regions.

In order to systematically document the salient features of environmental education projects, a system of project monitoring and evaluation is necessary. Monitoring and evaluation reports should document the progress of environmental education projects, and highlight the features of success, and failure; hopefully, the failures will be complimented by documented alternative plans. This type of information can be invaluable to planners of projects in environmental education. To see, for example, that a certain approach to awareness raising in Zambian rural areas was highly successful, can provide an essential base upon which new environmental education projects can be developed. Monitoring and evaluation reports can therefore be used as a 'blue print' for future environmental education projects.

Monitoring and evaluation not only has a retrospective value, it has the equally important function of providing the managers of environmental education projects with up-to-date reports of a project's progress - in terms of physical and financial terms, as well as the inputs and outputs of the project (as compared to the objectives of the project). In addition to giving the project managers (and other partisan groups) essential information of a project's development, it has the additional benefit of enhancing the likely chances of project funding. A donor who sees provision for monitoring and evaluation is likely to be slightly more biased towards supporting that project as compared to a project of similar merit that does not contain provisions for such activity. The provision for monitoring and evaluation will act to guarantee the donor with regular project up-date reports, and show that the project organiser is willing to adapt, and try new (alternative) techniques where it has been shown that a particular technique has been ineffective.

Monitoring and evaluation is not usually exclusively done by project staff. Some donor agencies have appointed their own in-country evaluation staff (e.g. Swedish International Development Agency) to undertake project evaluation. The evaluations in these instances tend to take place on a consultation basis. Project evaluation does involve a considerable degree of work, and can require the use of advanced research methods, data analysis etc. In light of this, it is unlikely that project staff will be responsible for undertaking project evaluations - unless it involves simple techniques. Additionally, post-project evaluations take place long after the project has ended, and the likelihood of project staff remaining in the area (or the implementing organisation) is slim. Monitoring, however, is essentially a management tool, which provides for information on project development, and allows for corrective action to be taken on the basis of the information provided. Monitoring involves the use of simple techniques, which do not require excessive time or effort on behalf of project staff to produce. The fundamental features of monitoring are: it is appropriate in terms of information, it is timely and it allows for corrective action. It is considered that in-service training should be able to provide key project staff with the appropriate skills to allow for project monitoring.

STEPS IN PROJECT EXECUTION

Planning and implementing an environmental education project is a complex and time-consuming task. In order to be efficient, there are a number of stages that should be considered. The following section examines a number of these stages although it is evident that the steps may differ, depending upon local circumstances.

An environmental education project cannot (or should not) be planned from an office far removed from the location of implementation. One of the merits of successful environmental education projects is that they are characterised by an on-going process of consultation between project organisers and the community (or beneficiary group). The consultative process can be lengthy, difficult, and often frustrating, but without imparting a sense of empowerment to the community, the project will not reach its objectives readily.

There are many considerations to be taken into account when planning an environmental education project - the majority of which must be addressed before submitting a project proposal. The time and effort required in planning, and then executing an environmental education project are considerable. Figures 1 - 3 (Leal Filho, 1994) lists some of these considerations, along with notes regarding their implications in the planning, implementation and post-execution of environmental education projects.

Table 1. *Elements to be considered in setting up an environmental education project.*

ELEMENT TO BE CONSIDERED	COMMENTS
STAGE I. BEGINNING	
1. Research local needs, capabilities and provisions	• Should be the community's perceived needs • Plan project within the scope of local conditions
2. Consideration to political, social, economic dimensions of scheme	• Project objectives should be commensurate with these factors - If not, what are the implications ?
3. Care with imposition of ideas and concepts	• Must select an appropriate approach • Ignoring local knowledge and skills will be at the organiser's peril !
4. Realistic budgeting	• Unrealistic budgeting shows a lack of planning and a level of incompetence - it will not impress a donor
5. Realistic timing	• Design time frame - it facilitates monitoring the progress of the project • Account for time lags / problems in time frame
6. Suitable staffing levels	• Exercise frugality when deciding on staffing levels • Staff salaries account for a larger proportion of a budget
7. Provisions for monitoring and evaluation	• Consider type of approach, time, staff required, and cost

EXAMPLES OF ENVIRONMENTAL EDUCATION PROJECTS

1. *'Environmental Camps for Conservation Awareness' (ECCA) - Nepal*

ECCA was established in 1987, and is a non-profit making organisation. ECCA has the primary objective of providing children with a broad environmental programme in the form of a five-day camp.

The project is mobile, to suit the topography of the country. Flexibility is essential in order that the project meets the needs of a culturally diverse population. Currently there are more than 40 camps, organised annually in over 18 districts.

Table 2. *Elements to be considered when running an environmental education project*

ELEMENT TO BE CONSIDERED	COMMENTS
STAGE II. DURING PROJECT IMPLEMENTATION	
1. Use of local expertise and input of local experts	• Use of indigenous knowledge and skills is more likely to result in greater chances of sustainability, and a lack of dependence upon outside help
2. Care with local human and logistical limitations	• Tailor expected results to fit with local human and logistical resources
3. Use of the environment as a teaching resource	• No substitute for the real environment • Communities will readily identify with their local environment
4. Deployment of monitoring / evaluation strategies	• Trial all instruments before use • Use a variety of methods (prevents inherent problem with one technique affecting the whole monitoring / evaluation

Table 3. *Elements to be considered after the execution of an environmental education project.*

ELEMENT TO BE CONSIDERED	COMMENTS
STAGE III. AFTER THE EXECUTION OF THE PROJECT	
1. Overall evaluation of tasks	• Consider all features of project, e.g. objectives, administration, staff performance
2. Dissemination of results	• Produce report and disseminate to all interested parties • Tailor reports for the differing parties
3. Post - execution measures	• Assess to what extent the recommendations of the monitoring / evaluation have been utilised

ECCA works in collaboration with a whole spectrum of grassroots, national and international NGOs, as well as government projects. The objectives of the project are as follows:

• To generate a sense of awareness amongst young people;
• To teach young people good outdoor habits, so that they may enjoy nature without adding to its deterioration;

- To provide insights into alternative technologies including energy, conservation, agriculture;
- To provide an opportunity to study in protected areas;
- To encourage young people to become involved in conservation.

ECCA avoids using formal methods of teaching in order to explicitly differentiate the activities of the project with that of formal education. It promotes the use of techniques such as encouraging learning through discovery and experimentation, and encouraging participation.

The project has been independently evaluated, and it has indicated a considerable level of success.

2. *'Chongololo and Conservation Clubs of Zambia' (CCCZ) - Zambia*

The Chongololo and Conservation Clubs of Zambia (CCCZ) is part of an initiative of the Wildlife Conservation Society of Zambia. The CCCZ began in 1972. It is organised as an out-of-school activity club, and the leaders are teachers. The clubs have developed a network, and primary ('Chongololo Clubs') and secondary schools (called 'Conservation Clubs') are a part of that network. An additional activity of the CCCZ is the production of a complementary radio programme. The objectives are:

- To promote efforts to conserve the natural diversity of Zambia's flora and fauna in balance with the needs of mankind;
- To develop environmental education projects and activities for the youth of Zambia and to support other organisations with the same aim;
- To provide a channel for the collection and distribution of conservation-related information.

The CCCZ produces newsletters, teacher's guides and the Chongololo magazine. The material suggests a mixture of theory, field and practical approaches. The Chongololo magazine is used as a stimulus for discussion and action. The theory aspects of the CCCZ involve the use of poems, music, drama, debates and experiments in order to develop awareness. The field work and practical sessions encourage familiarisation with a child's local environment, and gives an impression of how ecosystems function. The suggested activities are adjusted for the different ages of the children.

The Chongololo 'club of the air' is a weekly radio programme which supports the activities of the Magazines. The programme reaches young people who are not able to attend school. There is an English and Bemba-language version of the programme.

3. *'Enviroteach' - Namibia*

Enviroteach is an educational outreach project, formed by the Desert Research Foundation in Namibia. It is based at the Gobabeb research station, located in a desert region.

Enviroteach is a project which aims to meet the demand from teachers for materials in environmental education. Enviroteach designs and produces materials for secondary school teachers in all subjects.

The teaching materials are low-cost and high quality. The suitability of the materials is assured as all materials are produced in collaboration with the Ministry of Education and Culture. The objectives are:

- To design environmental education teaching materials to allow environmental education to be integrated into all subjects;
- To meet the specific demands of the teachers of Namibia;
- To disseminate the materials to all secondary schools in Namibia.

All new teaching materials are complemented by teacher training workshops, where Enviroteach staff introduce the materials, and discuss how the materials can be used in the classroom (and in practical lessons). At the end of each section of a manual, lesson ideas are provided for teachers in all subjects to encourage the introduction of environmental education into every subject.

Enviroteach pilot-tests all of its materials. This is done in 25 schools in the 4 educational regions of Namibia.

CONCLUSIONS

Why is a project approach important?

It is obvious that the use of environmental education projects should not be the exclusive method of introducing environmental education. The use of environmental education projects should be promoted for many reasons, including the following:

- Specific issues can be addressed, and thus an environmental education project becomes a problem-solving tool.

- As an environmental education project is tailored for a specific group of persons, it is likely to have a higher chance of success.
- There is a need for environmental education outside of the formal education sector.
- Projects provide an opportunity to work closely with a target group, to understand their role in the problem. They provide a platform for the exchange of ideas between the target group and educators.
- It is one of the approaches to environmental education which may produce readily tangible results, and possibly have a direct economic benefit.

REFERENCE

Leal Filho, W. (1994) 'Awareness raising and environmental projects in Less Developed Countries (LDCs)' In Analoui, F. (Ed) *The Realities of Managing Development Projects.* Avebury. UK.

Chapter 10

A CHECKLIST FOR THE CRITICAL EVALUATION OF INFORMAL ENVIRONMENTAL EDUCATION EXPERIENCES

by Roy R. Ballantyne & **David L. Uzzell**

School of Professional Studies
Queensland University of Technology
Locked Bag No. 2
Red Hill 4059
Australia

Department of Psychology
University of Surrey
Guildford
GU2 5XH
United Kingdom

INTRODUCTION

Since Belgrade and possibly earlier, teachers have perceived informal interpretive experiences as playing an important role in developing student environmental literacy. Visits to environmental centres, museums and heritage sites are undertaken to help students acquire environmental knowledge, concepts, skills, informed attitudes / values and sustainable environmental behaviour. The extent of this involvement is illustrated by the fact that the total volume of educational visits to heritage and tourism attractions (especially historic buildings, museums, art galleries, zoos and wildlife parks) in England in 1985 was estimated to be 12 million (Cooper and Latham, 1988). Of these, the majority of visitors were in the 9-13 years age group. Accordingly, it is not surprising that interpretive managers consider schools an important market, not only in their own right but also because they act as catalysts for future visits of families and friends.

THE NATURE OF INTERPRETIVE AND ENVIRONMENTAL EDUCATION LEARNING EXPERIENCES

Since the publication of Tilden's seminal book 'Interpreting Our Heritage' (1957), interpretation has been viewed as an important strategy to inform visitors to natural and cultural heritage sites about the uniqueness, value and fragility of different environments. Tilden defines interpretation as "an educational activity which aims to reveal meanings and relationships through the use of original objects, by first-hand experience and by illustrative media, rather than simply to communicate factual information". Aldridge (1972) extends this definition, making it prescriptive and action-oriented. He describes the purpose of interpretation as "... the art of explaining the place of man in his environment, to increase visitor or public awareness of the importance of this relationship, and to awaken a desire to contribute to environmental conservation". Interpretation is more than just providing people with facts about the environment. Rather, interpreters aim to make an object or event meaningful to the visitor by communicating its significance and sense of place or time in a way which relates to the visitor's own experience and personal world. Put simply, interpretation is the art of educating visitors of all ages by telling stories about the places they are visiting in an engaging, interesting and enjoyable way which, if successfully planned, leads to personal changes in environmental knowledge / concepts, skills, attitudes / values and behaviour. These aims are embodied in Tilden's principles of interpretation:

A. Any interpretation that does not somehow relate what is being displayed or described to something within the personality or experience of the visitor will be sterile;
B. Information, as such, is not interpretation. Interpretation is revelation based upon information. But they are entirely different things. However, all interpretation includes information;
C. Interpretation is an art, which combines many arts, whether the materials presented are scientific, historical or architectural. Any art is in some degree teachable;
D. The chief aim of interpretation is not instruction, but provocation;
E. Interpretation should aim to present a whole rather than a part, and must address itself to the whole man rather than any phase;
F. Interpretation addressed to children (say, up to the age of twelve) should not be a dilution of the presentation to adults, but should follow a fundamentally different approach. To be at its best it will require a separate programme.

While it is possible to discern similarities in the aims of environmental interpretation and environmental education, there are also important differences which have significant implications for their respective preparation, practice and evaluation. Both use educational principles to communicate environmental messages and develop environmental literacy. Environmental interpretation involves the use of informal learning experiences for people from a varied range of ages and abilities within a recreational setting. Environmental education on the other hand is principally directed at school students who are expected to develop environmental knowledge, skills, attitudes and behaviour through attendance at a combination of heuristic, classroom and field-based learning experiences. Advance preparation and subsequent assessment are important components of the formal education exercise. This is not the case with interpretation, where patrons - day visitors, family groups, tourists - are there by choice and seek an informative and entertaining experience. They will probably have undertaken no preparation for the visit and their time at a site may be limited. Any message, therefore, has to be communicated quickly and effectively. Finally, the casual visitor is not required to undertake any follow-up work in order to explore in detail the new material or reinforce the communicated message. This is not to say that the interpretation will neither inspire them nor be the beginning of a lifelong interest, but to expect substantive changes in people's perceptions and attitudes (which is an implicit goal of interpretation) after a twenty-minute encounter is to place hope before wisdom. There are, therefore, important differences between those who are visiting interpretive centres as part of an environmental education experience and those who are visiting as part of an interpretive experience. Hudson (1993) likens the difference to the distinction between conscripts and volunteers, with all the implications this has for motivation levels.

Research carried out in art galleries by Robinson (1928) over half a century ago demonstrates that many visitors spend only a few seconds looking at exhibits. Indeed, from the moment they enter a museum or art gallery, the amount of time spent in front of exhibits progressively decreases (Melton, 1972). This same phenomenon is found in interpretive centres. Behavioural tracking studies demonstrate that exhibits may be viewed on average for only a few seconds (de Vries Robbé, 1980). Accordingly, interpreters and exhibition designers create displays and experiences where meanings and relationships are communicated in fairly unambiguous ways and visitors are told what to believe. This contrasts with the heuristic approach more typically found in environmental education, where students are encouraged to discover, investigate and formulate principles and concepts for themselves. There is generally insufficient time for a day-trip visitor to become involved in such an in-depth, focused approach at an interpretive site.

General differences between environmental education and environmental interpretation are summarised in Figure 1 and can be analysed under the headings of WHO communicates WHAT to WHOM, WHY, WHERE, WHEN and HOW?

USING INFORMAL ENVIRONMENTAL LEARNING EXPERIENCES IN ENVIRONMENTAL EDUCATION

The incorporation of informal environmental learning experiences within an environmental education framework has many benefits. In particular, environmental learning occurs through direct exposure to objects and experiences rather than in a classroom setting. Students are able to apply theoretical knowledge in the field, discover real-life examples of environmental principles, problems and issues, and undertake problem-solving and environmental decision-making within a real world setting as well as being challenged to modify their environmental behaviour. Conventional and intuitive wisdom backed up by research (Fazio and Zanna, 1981) has shown that direct experience with an attitude object does lead to stronger attitudes compared with indirect experience. However, differences in design, content, target audience, purpose, educational setting, time constraints and educational approach have implications for teachers wishing to take students to interpretive sites and experiences. Teachers using environmental interpretive exhibits / experiences need to be aware of and make adjustments for mismatches in aims, target groups and educational methods. There are a number of issues, therefore, which need to be addressed when assessing the efficacy of informal environmental learning experiences with formal educational groups.

MATCHING INTERPRETIVE METHODOLOGIES TO THE AIMS OF FORMAL ENVIRONMENTAL EDUCATION

The aims and motivations which underlie interpretive displays / experiences are often broader than, and at times may even conflict with, formal learning aims. For example, the emphasis of interpretive displays / experiences may be on stimulating an affective or cathartic response or overtly encouraging attitude change which may be incompatible with curriculum or syllabus aims. Even when the aim is to promote learning, it will occur in an informal, recreational setting which may not be the most suitable, effective or desirable for visiting student groups.

It is also common to find that themes, stories and interpretive material are not organised in a sequence which enables students to incorporate easily new learning into cognitive frameworks previously developed in the classroom.

169

	ENVIRONMENTAL EDUCATION	ENVIRONMENTAL INTERPRETATION
Who?	School teachers; education officers; *teacher* educators	Interpreters; exhibition designers; scriptwriters; volunteers; archaeologists; education officers; academics; rangers; interpretive trainers
What?	Understanding of environmental concepts; acquisition of environmental skills; integration of environmental knowledge, attitudes and behaviour	Information about people, places, activities and objects; interpretation of meanings
To Whom?	Educational conscripts schoolchildren; student groups; adult/continuing education groups	Recreational volunteers: tourists; visitors; residents; wide age range
Why?	Develop environmental literacy; fulfil curriculum objectives	Recreational entertainment; profit; site conservation
Where?	Schools; field study centres; interpretive centres	Interpretive centres; historic houses; archaeological sites; urban and countryside sites; national, state and regional parks
When?	Timetabled periods; school trips; preparation and follow-up to visit	Whenever people engage in recreational or tourism activity; limited time involvement
How?	Instruction; heuristic and didactic techniques	Provocation; didactic and informal educational multi-media techniques

Table 1. Environmental education and environmental interpretation

Interpretive methods and techniques range from those that are passive and demand little from the learner to those augmented by sophisticated, interactive multi-media presentations. Interpretation, which demands the interactive participation of students, is clearly preferable in achieving the aims of environmental education, but such participation does not guarantee student learning. Techniques such as interactive audio-visual presentations, first-person interaction, re-enactments and sound / smell reconstructions which are found in many modern interpretive displays succeed in engaging students but may involve little more than the presentation and manipulation of factual information. To be of value in the achievement of environmental education aims, students should be involved in using higher order learning skills necessary for the development of environmental concepts, attitudes and behaviour.

MATCHING INTERPRETIVE EXPERIENCES TO THE AGE OF THE STUDENT GROUP

In most instances any exhibition and interpretive material will be aimed at a wide spectrum of visitors rather than focused upon the educational needs of young learners. Adults and children do not perceive the world in the same way and interpretive learning experiences aimed at adults may not achieve the intended effect with young students. Tilden (1957) suggests that interpretation addressed to children requires a fundamentally different approach. The discrepancy in the way adults and children experience and understand an interpretive display is confirmed by the research of Uzzell *et al.* (1984a; 1984b). Their investigation of responses to a cartoon-based interpretive exhibit indicates that, compared with adults, children appreciate different parts of the exhibition, value individual exhibits differently and have trouble understanding different exhibits. The evidence suggests that, in essence, adults and children 'see' different exhibitions. Accordingly, teachers need to consider the interests and abilities of their students and ensure that interpretive experiences are pitched at an appropriate level for their group.

MATCHING FACILITIES AND SERVICES TO THE NEEDS OF TEACHERS / STUDENTS

The needs of students and teachers during school visits to interpretive exhibits / experiences differ from those of recreationalists and tourists for whom the displays / experiences are chiefly designed. Uzzell and Parkin (1987) examined teachers' perceptions of the principal strengths and weaknesses of existing interpretive facilities in terms of their appropriateness for school visits. Where they were available, teachers listed as major strengths the provision of a 'real' learning experience such as the opportunity for practical and investigatory work; the opportunity to do more than just look and listen to content-oriented material; the provision of a variety of educational experiences (preferably indoors as well as outdoors); the availability of skilled interpretive staff with teaching experience; the provision of good teacher material and an awareness of current educational requirements; and a well-equipped classroom or workroom for student use. Such services are, however, available in only a small proportion of centres.

The principal failings of existing interpretive facilities for environmental education work were seen to be the inappropriateness of teacher and student material; untrained or inexperienced local staff; the overcrowding of facilities perhaps due to overbooking; oppressive restrictions on visitor activities; cheap and shoddy souvenirs; lack of wet-weather facilities (for eating brought or bought lunches) and workrooms; and interpretive facilities out of touch with the needs of schools and the background of students (particularly those from multi-cultural and ethnic areas). Uzzell and Parkin (1987) concluded that the facilities and services viewed as a priority for school visits are: the availability of both teacher and student education packs, specialist staff, lecture room facilities, price discounts (educational visits being very sensitive to price) and provision of interpretive material linked to national curriculum and national examination syllabi.

MATCHING INTERPRETIVE RESOURCES TO CONSTRAINTS OF NATIONAL CURRICULA

The impact of curriculum and institutional constraints on the undertaking of environmental fieldwork will vary from one country to another. In Britain there has been a trend for the centralisation of the school curriculum by the former Department of Education, and a reduction in teacher autonomy as to what is taught in the classroom. The National Curriculum lays down tight guidelines as to what should be taught to pupils at different ages, backed up by attainment targets at 'key stages'. Forward thinking interpretive centres and museums, recognising the importance of the schools market, not only provide exhibitions with this in mind but also provide teacher and student resource materials and other curriculum-related educational materials in order to encourage school visits. Teachers have neither the time nor the resources to stray far from the National Curriculum. Any educational investment on the teacher's part is assessed in terms of the costs and benefits of a school visit.

While it may be argued that not all school group visits are strongly curriculum driven, research indicates that the majority do have a curriculum-related purpose. For instance, Keeley (1992) found that 58% of primary school (7-11 years) and 53% of secondary school (12-18 years) visits to attractions were organised in support of specific curriculum activities, while 28% (primary) and 24% (secondary) of visits were for general educational and development benefit. Finally, 12% (primary) and 21% (secondary) had no specifiable educational intention at all, being for social and recreational purposes.

EVALUATING AND ALTERING INFORMAL ENVIRONMENTAL LEARNING EXPERIENCES

In the light of the issues discussed above, teachers wishing to access interpretive displays / experiences to develop student environmental literacy need to evaluate and, if necessary, alter the interpretive learning situation to meet their own requirements. Similarly, interpreters may wish to evaluate and alter their displays in order to attract the formal education market. To this end, both teachers and interpreters need access to evaluative and prescriptive tools which focus on the use of informal environmental learning experiences within an environmental education context. The Informal Environmental Learning (IEL) checklist has been developed to meet this need.

CHECKLISTS AS AN EVALUATIVE TOOL

A checklist consists of a series of statements which "describe the critical attributes of either some procedure or product" and are particularly useful in providing "the basis for evaluation of the performance or outcome" (Geis, 1984). They include "the qualities of actions necessary in a desired performance or the qualities of elements necessary in a desired product" (Yelon, 1984). Although often perceived as static (the image of a person mindlessly ticking the existence or absence of items comes to mind), checklists can be dynamic, interactive and play an important role as a teaching / learning aid by promoting the critical evaluation of a performance or a product. Of relevance to the needs of environmental educators and interpreters is their value in providing a time and cost-effective way of evaluating and improving informal environmental interpretive learning experiences.

According to Yelon (1984), there are four circumstances in which the benefits gained by the use of checklists are sufficient to justify the time and effort spent in their creation and use. These are when a performance is a very important competency upon which others depend; when a skill is particularly complicated and it is important to check each part of the performance; when precise feedback is needed to communicate needed improvements; and when independent, self-directed learning is required. Elements of all these conditions are relevant in the context of evaluating informal environmental learning experiences. Of particular importance is the need for a self-administered instrument which enables the user to identify specific educational shortcomings within an often complex display/experience and plan steps to be taken to improve the learning experience. According to Yelon's criteria, the use of a checklist is appropriate and likely to be highly effective for this purpose.

Silverstone (1989) cautions that "the heritage industry is in the business of mass communication and the boundary between museums and media and that between reality and fantasy, between myth and mimesis in both sets of institutions and practices is becoming increasingly blurred, increasingly indistinct". It is precisely because the heritage and environmental industry have such difficulties differentiating between fact and fiction that teachers and interpreters need a rigorous and systematic set of criteria to help judge as objectively as possible the potential value of an environmental display or experience. Although a checklist may not enable judgements to be made about integrity and authenticity, it can play an important role by alerting individuals to the dangers of being swept along by the medium rather than the message.

THE INFORMAL ENVIRONMENTAL LEARNING CHECKLIST

The Informal Environmental Learning (IEL) checklist is designed to be used by environmental interpreters and teachers to promote the critical evaluation, and thereby enhance the effectiveness, of informal environmental experiences for student learning. Alternatively, student interpreters and teachers could be involved in creating their own checklists as a learning exercise (Geis, 1984), using the IEL checklist as an exemplar against which individuals can compare, debate and evaluate their own criteria.

The value of using a checklist to assess interpretive experiences was initially recognised by the National Awards Panel of the Carnegie Interpret Britain Awards Scheme. An earlier version of the checklist was developed by one of the present authors and members of the Panel as part of the assessment procedures of the Awards Scheme for good interpretive practice organised by the Society for the Interpretation of Britain's Heritage. Using the Interpret Britain Awards Scheme checklist as a starting point the IEL checklist was developed, trialed and refined with the assistance of interpreters and teachers in both the United Kingdom and Australia. The IEL checklist can be used by teachers for evaluating and improving informal environmental learning experiences for students; by practitioners and higher education environmental educators and interpreters in the pre- and in-service training of teachers and interpretive staff; and by managers of environmental displays and experiences to evaluate the success of these in meeting the needs of school groups. The IEL checklist is reproduced below.

THE INFORMAL ENVIRONMENTAL LEARNING (IEL) CHECKLIST

1. *Pre-visit preparatory requirements*
1.1 Are there courses or induction programmes available for teachers prior to a visit?
1.2 Are staff available to discuss ways of maximising student environmental learning prior to their visit?
1.3 Are there pre-interpretive visit information brochures, education packs and resource materials for students and / or teachers?
1.4 Is this material interdisciplinary?
1.5 How is the material accessed and is there an extra cost?

2. *The learning environment*
2.1 Is the physical environment conducive to learning?
2.2 Is the social environment conducive to learning?
2.3 Does site management display imagination and flexibility in dealing with student groups?
2.4 Are there classroom / workroom / wet weather facilities?

175

2.5 Is there an interesting and stimulating programme of events and other activities which support environmental learning?

2.6 Are exemplary practices modelled, i.e. waste management, energy conservation?

2.7 Are there a variety of educational experiences and resources on offer both indoors and outdoors?

2.8 Will students with disabilities be catered for, e.g. mobility, visually handicapped?

2.9 Are educational groups made to feel welcome?

3. *Appropriateness of educational material*

3.1 Is there a specific environmental education programme or programmes?

3.2 Do the themes, messages, stories and information relate to the experience and world of students?

3.3 Do the themes, messages, stories and information reinforce or exemplify concepts developed in the classroom?

3.4 Is the educational provision linked to school curricula and syllabi?

3.5 To what extent does the interpretation tend to stimulate understanding and awareness of environmental concepts, attitudes and appropriate behaviour?

3.6 Are the staff sensitive to the cultural and ethnic background of students in the design of educational and interpretive materials and programmes? Is material gender inclusive?

3.7 Are there a variety of educational learning experiences on offer suitable for the particular student age group?

3.8 Are environmental themes and stories linked within an appropriate framework?

3.9 Is provision made for varying levels of student understanding and interest?

3.10 Are the ideas and factual material accurate and up-to-date?

3.11 Are the topics and themes appropriate to the site being interpreted?

3.12 What are the overriding messages of the display / experience? Are they clear and compatible?

4. *Learning techniques*

4.1 Does the interpretation provide the opportunity for student involvement in environmental decision-making and problem-solving?

4.2 Do students appear to be motivated by the displays / experiences to engage in self-directed learning?

4.3 Are there opportunities for practical and investigatory work?

4.4 Does the educational material / programme lead to the acquisition of skills as well as information?

4.5 Is material sequenced to help promote the integration of learning?

5. *Design*

5.1 Are materials designed by individuals with teaching experience and awareness of current educational requirements?

5.2 How appropriate are the selected media for the setting and circumstances; do they appear incongruous in the particular setting?

5.3 Is the technical standard of presentation high in respect of the texts, photographs, drawings and diagrams?

5.4 Are the displays legible, conveniently visible and pitched at an appropriate reading level?

5.5 Are the topics and themes pursued with appropriate thoroughness in the presentations?

5.6 Does the interpretation / education material attract and hold the attention of students?

5.7 Is there flexibility in arrangement / mobility of displays to meet particular group needs?

5.8 Are different learning styles catered for, e.g. audio, visual, kinaesthetic?

6. *Follow-up*

6.1 Are there post-interpretive visit education packs and resource materials for students and / or teachers? If there is a cost, is it 'value for money'?

6.2 Is there any follow-up such as literature to take away, environmental groups to join or suggestions of other places to visit that would extend the interpretation?

6.3 Does the facility / service appear to generate lasting interest which may lead students to pursue further the subject interpreted?

6.4 Are there suggestions of actions which could be undertaken as a result of the environmental learning experience?

6.5 Are there good examples of student follow-up work on display?

7. *Staff performance*

7.1 Is there a good basic staff structure to provide the appropriate services to student visitors, e.g. education officer?

7.2 Does the demeanour and communicativeness of the staff enhance student experience, understanding and enjoyment of the display / experience?

7.3 Do the staff appear to know their facility, its history, activities and services?

7.4 Is there evidence of staff training programmes which deal with the learning needs of students?

8. Management

8.1 Is all hardware functioning properly, e.g. visual aids, interactive media?

8.2 Is there evidence of systematic evaluation in order to keep abreast of teacher / student needs and subsequent modification in the light of this evaluation?

8.3 Is the site clean and well cared for?

8.4 Is the site safe and hazard free?

8.5 Are there approved procedures for dealing with emergencies (e.g. injuries)?

8.6 Will you receive all the attention, facilities and services you require, or will you be competing for resources with other groups booked at the same time?

8.7 Is there any risk that your student group will come into conflict with tourist groups or others, to the disadvantage of both groups?

9. Other

9.1 Are there discounts for student groups? Are accompanying / supervising adults charged?

9.2 Do the facilities, educational experiences and services appear to be good value for money?

9.3 Is there adequate, safe and convenient parking for buses and cars?

9.4 Is there a meeting area to address a large group of students?

9.5 Is there an area for the storing of school bags?

9.6 Are the refreshment facilities adequate?

9.7 Are the toilet facilities adequate?

9.8 If required, are there suitable and good value souvenirs available for purchase?

9.9 Are there facilities (under cover) for eating bought or brought lunches?

9.10 If appropriate, are there overnight facilities available on site or nearby?

CONCLUSION

How should the IEL checklist be used? It can be used as an *aide-memoire* in an informal way. Used in this manner it will provide a structured means of thinking about each stage of an educational visit to a museum or interpretive site, from preparatory work, through to site visit and follow-up. Equally, it can be used in a more systematic and quantitative way. The checklist has been sub-divided into nine sections. Each question can be assigned a point. It may then be decided that it is necessary for a facility to achieve at least 50% for each section. Some sections may also be double weighted because of their importance. As a result, a threshold score can be derived for the whole checklist: an interpretive facility would only be considered worthy of a visit if it reached this threshold. Such a scoring system may seem overly prescriptive, but it does encourage the environmental education course leader to assess in a more rigorous and systematic way the efficacy of a visit to a museum or site. It may also help to reassure those whom he or she has to convince that the choice of site has been evaluated in a professional, comprehensive and demanding way.

When should the IEL checklist be used? Teachers should, of course, visit all museums or interpretive sites and centres before taking parties of school children to them. Indeed, many facilities make this a requirement if the facility is to provide any services such as a talk by an Education Officer. This serves the purpose not only of preparing the teachers and allowing them to become acquainted with the staff, resources and site, but also of enabling them to discuss their requirements with the centre or museum staff. The IEL checklist should be an important part of the pre-visit preparations. It may indicate problems in the design and management of informal environmental learning experiences. The results can be discussed with the centre / museum in order to see whether any of the shortcomings can be addressed.

Who should use the IEL checklist? Clearly, any teacher or co-ordinator of environmental education courses who uses museums, heritage sites or any other informal educational facility to augment teaching and learning in the classroom will benefit from use of the checklist. Equally, the checklist will be of use to providers of environmental education and interpretive facilities and services. Do they meet the kind of demanding criteria that course organisers should and are now making? What aspects of the service they provide are deficient? What aspects are exceptional and should be promoted to potential school and student groups? The IEL checklist will help identify the strengths and weaknesses of current provision and suggest opportunities for further enhancement and improvement. In this way, the quality of informal environmental learning experiences will be maximised and their use in formal environmental education programmes more soundly justified.

REFERENCES

Aldridge, D. (1972) 'Upgrading Park Interpretation and Communication with the Public'. Paper presented at *Second World Conference on National Parks*, Yellowstone, USA.

Cooper, C.P. and Latham, J. (1988) 'The Pattern of Educational Visits in England'. *Journal Leisure Studies*, Vol 7, Part 3, pp. 255-266.

De Vries Robbé, G. (1980) *Countryside Interpretation: the Interaction between Countryside Visitors and Interpretive Displays*. Unpublished M.Sc. Dissertation, University of Surrey.

Fazio, R. and Zanna, M. (1981) 'Direct experience and attitude-behaviour consistency'. *Advances in Experimental Social Psychology*. Vol. 14, pp. 161-202.

Geis, G.L. (1984) 'Checklisting'. *Journal of Instructional Development*. Vol. 7, No.1, pp. 2-9.

Hudson, K. (1993) 'Visitor Studies: luxuries, placebos, or useful tools?' In S. Bicknell and G. Farmelo (Eds.) *Museum Visitor Studies in the '90s*. Science Museum, London.

Keeley, P. (1992) 'The School Visits Market' *Insights*. Vol. 5 A133-139.

Melton, E. (1972) 'Visitor behaviour in museums: some early research in environmental design'. *Journal Human Factors*. Vol. 14, No. 5, pp. 393-403.

Robinson, E.S. (1928) '*The Behaviour of the Museum Visitor*'. American Association of Museums: New Series, No.5, Washington, D.C. Silverstone, R. (1989) 'Heritage as media: some implications for research' In D.L. Uzzell (Ed.) *Heritage Interpretation, Volume II: The Visitor Experience*. Belhaven Press. London.

Tilden, F. (1957) '*Interpreting our Heritage*'. University of North Carolina Press, Chapel Hill, USA.

Uzzell, D.L., Lee, T.R. and Henderson, J. (1984a) '*Interpretive Provision in Forestry Commission Visitor Centres: Strathyre Visitor Centre*'. Report to the Forestry Commission, University of Surrey.

Uzzell, D.L., Lee, T.R. and Henderson, J. (1984b) '*Children at the Strathyre Visitor Centre*'. Report to the Forestry Commission, University of Surrey.

Uzzell, D.L. and Parkin, I.C.A. (1987) '*Bedgebury National Pinetum and Forest: Interpretive Plan*'. Report to the Forestry Commission, Edinburgh.

Yelon, S.L. (1984) 'How to use and create criterion checklists' *Performance and Instruction*. Vol. 23, No. 3, pp. 1-4. Originally featured in:

A checklist for the critical evaluation of informal environmental learning
experiences

International Journal of Environmental Education &Information.
Vol. 13, No. 2 1994 (published by Environmental Resources Unit,
University of Salford, Salford MS 4WT, U.K.)

Chapter 11

FURTHERING ENVIRONMENTAL EDUCATION

Walter Leal Filho

European Research and Training Centre on Environmental Education
University of Bradford
Department of Environmental Science
West Yorkshire UK
BD7 1DP

INTRODUCTION

As seen throughout this book, there is wide acceptance of the fact that the Belgrade Workshop was one of the benchmarks in the historical evolution of environmental education. Various significant developments in international environmental education have taken place since, along with a number of significant progressions within countries.

If environmental education today is seen as a well established methodology (Leal Filho, 1995a) and if individual countries - especially the developing ones - attach some relevance to it, this is partially due to the ideas set in motion at Belgrade, to the establishment of the International Environmental Education Programme (IEEP) and to the extensive work executed since then. The earlier sections in this book have reviewed the mechanisms set in motion at Belgrade (1975), IEEP's establishment and the impact of the meetings in Tbilisi (1977), Moscow (1987) and of the UN Conference on Environment and Development (UNCED), held in Rio de Janeiro, Brazil in 1992. It is now relevant to discuss some key elements that characterise modern environmental education. These are:

- environmental education and sustainable development
- environmental education and biodiversity conservation
- environmental education and indigenous knowledge

Following this, some critical ideas will be presented, outlining some of the items that need to be taken into account when further developing environmental education.

ENVIRONMENTAL EDUCATION AND SUSTAINABLE DEVELOPMENT

There is no doubt that a close link exists between environmental education and a country's ability to reach the ultimate goal of sustainable development. In view of the relevance and the potential of this relationship, numerous efforts have been made over the past two decades to promote environmental considerations in the context of economic development programmes (ICDI, 1980, Meadows, 1990). Particular momentum was gained with the preparation of the report 'Our Common Future' (WCED, 1987), which advocated debate at the highest level on the topic. Further inputs were gathered from UNCED, which approved a number of measures aimed at the integration of economic development with environmental conservation.

A noticeable degree of progress was seen during the UNCED process, with various country reports, describing the state of the environment in each nation, being prepared. The fact that national reports on environment and development issues were prepared - including in countries known for their low priority awarded to environmental matters - illustrates the fact that sometimes the process is more important than the result.

Leal Filho (1995a) has presented a number of definitions used by governments in developing countries to describe sustainable development. Some of these are:

* a model of development that takes into account a country's need for wisely using its environmental resources;
* an approach to development that limits the impact of human activities on the environment;
* a methodology of development that foresees some limited damages to the environment, but for the general good;
* a type of development that cares for the environment but at the same time shows awareness of the need for the sound use of natural resources to achieve progress.

No matter the definition that one favours, the underlying features of the relationship between environmental education and sustainable development rely on the fact that:

* environmental education, seen in the context of education for the environment (Fien, 1992) is concerned with an improvement in the relationship between human beings and the environment, which

            includes - but is not limited to - reductions in the level of environmental degradation;

*   environmental education includes in its context the social, economic and social issues that permeate the environment, including matters such as poverty alleviation, hunger (related to food scarcity) and essential health topics such as parasitic diseases such as 'ascaridiasis' - common across the developing world - which directly reflect on productivity at work and thus slows down a country's capacity to produce goods or crops;

*   environmental education is concerned with activities which may take place in schools, as advocated by Huckle (1990), Tilbury (1992) and Fien (1993), but also at community level as stated by Haider (1992), Orr (1992) and Novo (1993), bringing the discussion on environmental matters to a wider audience.

As development is essentially a process of structural transformation in a society, in which resources are made available - more evenly - to more people, it can be seen that environmental education is not only a process compatible with its aims, but also an effective tool towards reaching that goal: that is, the sustainable use of environmental resources.

    It needs to be recognised that environmental education initiatives which claim to work towards the goal of sustainable development do not always deal with issues of key relevance to developing countries, although matters such as climate change or ozone depletion (Boyes, *et al.*, 1995; Leal Filho, 1995b) are important items. Some of the issues which are not normally tackled, despite their real and immediate value to people in developing nations are:

i. malnutrition;
ii. poverty;
iii. alternative crop production;
iv. energy efficiency;
v. renewable fuel sources.

Although the above list is by no means comprehensive, it can be seen that part of the root of the problem is that there is little interest - and often little expertise - from donor or international agencies to ensure that environmental education is applied when dealing with these issues. This state of affairs illustrates the relevance of, and the need for, creating local capacity among people in developing countries, to critically look at and deal with these problems. The goal of economically sound sustainable growth and the improvement of living conditions cannot possibly be achieved by developing countries if they are not able to, on their own, find solutions to the ecological, as well as social, political and economic problems that exist within their borders. If it is to succeed, environmental education, particularly in developing nations, must take into account such a context.

ENVIRONMENTAL EDUCATION APPLIED TO BIODIVERSITY CONSERVATION

In addition to the usefulness of environmental education as a tool for sustainable development in the general sense of the word, there is a need to consider an aspect related to it, which applies to the conservation of plant and animal species. It is necessary to consider, in other words, environmental education applied to biodiversity conservation.

There is a consensus among environmentalists and economists in relation to the fact that the current model of development adopted in the majority of countries has been, in most cases, based on the heavy exploitation of natural resources, with a subsequent negative impact on a nation's fauna and flora. Although the impact of environmental degradation is felt with different levels of intensity among countries, there are clear examples where environmental degradation has been leading to losses in a country's plant and animal species, or in other words, in a loss of biodiversity.

On the basis of the need to tackle the issues which are related to environmental degradation, and which have an impact on both plants and animals, there has been a noticeable re-orientation of some environmental education programmes towards addressing these issues. Environmental education applied to biodiversity conservation, also called by some as 'education for biodiversity', represents the set of strategies, techniques and approaches which look at the issues of biodiversity conservation and the protection of wildlife (plants and animals included) with an educational angle.

There have been, over the past few years, significant developments in global attempts to apply principles and practices of environmental education to biodiversity conservation programmes. International organisations such as IUCN and WWF, as well as the World Bank's Global Environmental Facility and the development authorities of the UK (ODA), US (USAID), Norway (NORAD), Sweden (SIDA) and many others, are supporting a wide range of projects in biodiversity conservation.

A key feature of biodiversity conservation programmes is the fact that, implicit in them, is a set of educational aspects. This line of thinking is endorsed and indeed included in the 'Convention on Biological Diversity'. Article 13 of the Convention, which focuses on 'Public Education and Awareness', states that "contracting parties shall:

a) promote and encourage understanding of the importance of, and the measures required for, the conservation of biological diversity, as well as its propagation through media, and the inclusion of these topics in educational programmes; and

b) co-operate, as appropriate, with other States and international organisations in developing educational and public awareness programmes, with respect to conservation and sustainable use of biological diversity."

In the context of the above calls to promote awareness of biodiversity and its conservation, environmental education can play a key role, hence the need for including an awareness-raising dimension in the context of biodiversity projects. Environmental education applied to biodiversity conservation, (as it is the case of environmental education techniques targeted at reaching the goal of sustainable development), is inter-related. The conservation of a country's biodiversity is a pre-condition for it to develop itself sustainably. Whenever plant or animal species are destroyed, a proportion of a country's wealth is being destroyed with them, which makes that process unsustainable. Within the universe of biodiversity conservation activities seen across the world, it is noticeable that a substantial proportion of them are giving due consideration to the need for combining conservation measures, with educational measures aimed at integrating them with the other social aspects of the conservation process. In order to successfully deal with all the relevant issues, biodiversity conservation projects need to have - implicitly or explicitly - an environmental education component.

THE ROLE OF INDIGENOUS KNOWLEDGE

The story is told of an Indian tribe in South America. An energy company decided to establish a hydroelectric power station in the area in which the tribe lived. The setting-up of the station would not only imply the re-allocation of the tribe to another area, but also in the flooding of several square kilometres of forest land, with the subsequent destruction, by the water, of the plant and animal species found there. Although extensive studies on the feasibility of the scheme were performed, no consultation with the local tribe was made.

One day, when informed by a government official that they would need to move elsewhere as a dam was to be built in the site, the chief of the tribe said: "only a fool would build any big buildings here". When asked why by the government official, the chief replied that the site was known as "shaky land" in his own idiom, due to the fact that the unstable soil would not withstand any large construction. Concerned with the comments made by the chief, the local government performed a geological study which indeed pointed out details in the nature of the soil and its unsuitability for large construction works, which were overlooked by previous feasibility studies.

The example above, although simplistic, illustrates the relevance of taking into account local knowledge. In the environmental field, there are various records of situations when the provision of local knowledge has helped projects to be successful, and ideas to become reality. Modern medicine has also benefited a great deal: medicines now commonly used (e.g. curare, used as an anaesthetic) only came into commercial use due to the fact that information about it was passed on by Amazon Indians.

The same line of thinking also applies to environmental education. Indigenous knowledge on the environment exists and is seen as a very important tool in fostering environmental learning. As stated by Lalonde and Akhtar (1994), "over the last few years a consensus has emerged in the scientific literature and through various international gatherings on the legitimate field of environmental expertise known variously as indigenous knowledge (IK) or indigenous environmental knowledge (IEK)". However, although on the one hand indigenous knowledge may add extra value to environmental education initiatives, there are, in relative terms, few initiatives in this field. This state of affairs is seen today as a result of a wide range of factors. Some of these are:

i)   in many countries, indigenous knowledge is often seen as an issue of 'folklore value', rather than as an efficient, reliable set of information which can greatly help people to deal with environmental problems;

ii)   in institutional terms, there are few organisations which may systematically draw on the experiences seen in indigenous knowledge. As a result, most of the potential in this field is untapped;

iii)   few of the experiences gathered through indigenous knowledge are properly documented. It is difficult to find documentation on ways whereby indigenous knowledge can be used, for the simple reason that experiences in this area are not always recorded. Lack of documentation has various implications, particularly on matters related to how to use the knowledge and how to disseminate information arising from indigenous people.

Despite the above, indigenous knowledge can be of great help towards raising environmental awareness in a number of ways. Firstly, because it uses information locally available, whose value is based on an established record of testing and use (e.g. use of medicinal plants). Secondly, because it brings along concepts related to the best ways whereby environmental resources may be used: this often implies a minimum environmental impact (e.g. the collection of fallen branches for fuelwood). Thirdly, because indigenous knowledge relates to a deep insight into environmental matters and phenomena at a local level, which may enable well-informed decisions to be made (e.g. the best time of the year to plant new crops).

It is thus seen that indigenous knowledge is an important aspect of the social sciences, which may be of direct use to environmental education. The benefits that can be derived from it also include better information about the environment, and its capability to catalyse better behaviour in relation to the ways in which environmental resources are used.

EDUCATION FOR SUSTAINABILITY OR EDUCATION FOR BIODIVERSITY?

The last few years have seen the development of two fairly new and similar expressions, which refer to two of the goals of environmental education previously described in this chapter, and which are related to the promotion of sustainability and the conservation of biodiversity. These expressions, namely 'education for sustainability' (EFS) and 'education for biodiversity' (EFB) have been influenced by recent publications such as 'Caring for the Earth' (UNEP, IUCN, WWF, 1990) and others. They are also complemented by expressions such as 'education for sustainable development' (EFSD), 'environment and development education' (EDE) and older expressions which date from the late 1970s and early 1980s such as 'conservation education' (CE).

Although on the one hand the use of such expressions may reflect an individual's way of seeing environmental education, and may ultimately contribute towards the promotion of environmental education, concerns about both the multiplicity and the meaning of such expressions and the negative impact they might have outside the context of industrialised countries is growing.

Based on extensive field experience, the author advocates the view that the diversity of expressions currently used, all of which - it should be stressed - bear a direct link with environmental education, may confuse (rather than clarify) its relevance. This state of affairs is particularly worrying in developing countries for two main reasons:

i. there is virtually no equivalent expressions in idioms other than English which can fully describe their meaning without a detailed explanation of what they in fact mean;

ii. in many developing nations, the very concept of environmental education is still at an embryonic stage. In such circumstances, the insertion of new terms may be negative to attempts, over the past 20 years, to clarify what 'environmental education' in fact means.

As can be seen in Figure 1, the wide range of terms currently used originated from, and tend to converge in, environmental education. It is thus important that attempts to promote terms such as EFS, EFB, EFSD, EDE and others are based on careful reflection and on a critical analysis of the extent to which such terms serve the purpose of furthering environmental education or whether they will fragment it into so many small pieces that the view of the whole becomes compromised.

Figure 1. The applied nature of environmental education

A common feature among these expressions is that they all confirm the applied nature of environmental education and illustrate its applicability to various contexts. Independent of the term one favours, the real challenge is to ensure that

189

environmental education is incorporated as part of teaching systems - as well as in the context of non-formal programmes - and that an environmental awareness dimension is promoted as part of economic activities, all of which, to a lesser or a greater extent, have an impact on the environment. Over and above these challenges is the need to clarify what the environment is, how people's behaviour affect it (Newhouse, 1991; Scott and Willits, 1994) and to motivate people to behave in a way that is compatible with the conditions of the environment in which they live.

These goals can only be fully reached if environmental education is part of the teaching system (Stapp *et al.*, 1980), and seen in a global context (Porritt, 1991; Boulding, 1988).

MOVING AHEAD

Due to the widespread use of environmental education as a methodology aimed at fostering environmental conservation, and the fact that it has a number of characteristics which make it multi-applicable, it is reasonable to expect that its growth in terms of use will be accompanied by improvements in the variety of methods, approaches and initiatives in both formal and non-formal teaching. Although twenty years have passed since the Belgrade Workshop and since UNESCO-UNEP's International Environmental Education Programme was created, some fundamental questions related to environmental education still remain. Some of them are:

1. Which characteristics should education concerning the environment have?
2. How to introduce environmental education into the pool of teaching practices?
3. At which stage of the educational process should environmental education be incorporated?
4. Which are the strategies (and materials) most adequate for action, in formal teaching, in this field?

The answer to these questions is not easy, although, as described elsewhere, a great deal of progress has been made in attempts to address them (e.g. National Milieubeleidsplan, 1989; UNESCO-UNEP, 1990; CEE, 1994; Leal Filho, 1994a; Leal Filho 1994b). However, it needs to be said that the systematic development of environmental education can only be fully achieved if - in addition to the answers to these questions - a number of requirements (both structural and logistical) are fulfilled and, as stated by Beare and Slaughter (1992) and Hicks and Holden (1995) future possibilities are explored. Some of these items, which are particularly relevant in a developing country's context are the six 'Ps'. These are: philosophy, policy, plan of action, programmes, participation and partnerships. Due to their relevance, they are described in turn.

Philosophy: individual countries need to adopt a philosophy on environmental education. This needs to be based on what they perceive the 'environment' to be and how they perceive environmental education. The value system applied will be reflected in the policies and the methods used to promote environmental education in both formal and non-formal teaching.

Policy: individual countries should establish clear policies, programmes, plans or guidelines on environmental education. Independent of the terminology used, it is vital that a country clearly states what it intends to do in this field, how it intends to do it and when it intends to do it. Through this procedure, parameters are established, against which progress may be checked.

Plan of Action: a policy on environmental education needs to be supplemented by a plan of action, which outlines the various tasks to be performed over a determined period of time. Plans of action also facilitate the division of tasks.

Programmes: plans of action should be structured around programmes, which in their turn may consist of projects. Programmes will have a beginning, a middle and an end, during and after which results may be assessed.

Participation: environmental education cannot be practised without participation. Public participation in environmental issues is essential to ensure long-term results. There is on the other hand a perceived imbalance in the traditional focus of environmental education which needs to be tackled with a view to fostering more participation. Most of the current initiatives focus on formal teaching, while non-formal environmental education programmes or those which have a community focus, are still in a minority. If participation in environmental education is desirable and if it is to be made accessible to all people, it will need to be more evenly spread, also reaching those groups which are not reached by formal schooling (as well as those who do).

Partnerships: the popular saying that 'no man is an island' does apply to environmental education and is particularly relevant if we consider the fact that skills are found spread over various individuals and institutions. However, a partnership is not a one-way process. There needs to be an authentic interest, among the concerned parties, to complement each other's efforts and respect the various views, experiences and ideological positions of the partners. Unfortunately, despite the great benefits to be reaped through partnerships, not enough emphasis is being given to them in the field of environmental education.

These items, which are by no means the only ones that need to be tackled, ought to be carefully considered as they are behind some of the barriers which have been preventing the development of environmental education in both industrialised and developing countries, since Belgrade and even earlier.

Another item that needs to be looked at is the issue of documentation of environmental education initiatives (e.g. projects or programmes). Unfortunately, not all experiences in this field are properly documented, which allows overlap and duplication. This specific matter is yet to be fully dealt with, although attempts to document initiatives have been made (e.g. British Council, 1993; Leal Filho, MacDermott and Murphy, 1995).

Last but not least, there is a need, especially among researchers and practitioners, to integrate their approaches and practices and identify ways in which they can work together. Research on environmental education has evolved dramatically over the past few years, especially in Britain and in Europe (Leal Filho and McDowell, 1995; Leal Filho and MacDermott, 1995). Data gathered from research can be of great value to practitioners.

CONCLUSIONS

The world is constantly evolving and so should environmental education as well as the ways in which it is seen. The Tbilisi Conference endorsed Bloom's Taxonomy as the best method for describing environmental education goals. It is now seen that new approaches are needed to complement it and maximise the action-oriented dimension of environmental education.

Countries need to make a stand as to which issues they wish to deal with, and which priorities they ought to have. By following this procedure, work on the ground ceases to be donor-driven (hence unsustainable) and becomes locally-focused, with a greater likelihood of success and possibilities of continuation when a donor's engagement is terminated. In doing so, we can ensure that environmental education is furthered and that initiatives in this field cease to be isolated and, instead, become sustainable.

Although there is no such a thing as the 'ideal' approach to environmental education, it is believed that, to further environmental education, it is important to consider it as a cross-cutting issue, in which biodiversity conservation is an important component, for which indigenous knowledge is taken into account, and through which the goal of sustainable development may be achieved.

REFERENCES

Beare, H. & Slaughter, R. (1992) *'Education for the Twenty-First Century'*. Routledge, London.

Boulding, E. (1988) *Building a Global Civic Culture: Education for an Interdependent world.* Teacher College Press, New York.

Boyes, E., Chambers W. & Stanisstreet M. (1995) 'Trainee primary teachers' ideas about the ozone layer' *In Environmental Education Research.* Vol. 1 (2), pp. 133-145.

British Council (1993) *Environmental Education in Schools. Report of the Seminar held at the British Council.* 16 March 1993, The British Council, Manchester.

Fien, J. (1992) *'Education for the Environment'.* Unpublished doctoral thesis. University of Queensland, Brisbane.

Fien, J. (1993) *Education for the Environment - Critical Curriculum Theorising and Environmental Education.* Deakin University Press, Geelong.

Haider, M. (1992) *Umeltgestaltung fur die Zukunft: Prinzipien und Uberzeugende Strategien.* Quintessenz-Verlag. Munich.

Hicks, D. & Holden, C. (1995) 'Exploring the future: a missing dimension in environmental education', *In Environmental Education Research.* Vol. 1 (2), pp. 185-193.

Huckle, J. (1990) 'Environmental education: teaching for a sustainable future', In B Dufour (Ed) *The New Social Curriculum.* Cambridge University Press, Cambridge.

ICDI (1980) *North-South: a programme for survival: The Brandt Report.* Pan Books, London.

Lalonde, A. & Akhtar, S. (1994) 'Traditional knowledge research for sustainable development', In *Nature & Resources.* Vol. 30 (2), pp. 22-28.

Leal Filho, W.D.S. (Ed) (1994a) *Trends in Environmental Education in Europe.* ERTCEE, Bradford.

Leal Filho, W.D.S. (Ed) (1994b) *Environmental Education in Small Island Developing States.* Commonwealth of Learning, Vancouver.

Leal Filho, W.D.S. & McDowell, J. (Eds.) (1995) *Proceedings of the National Seminar on Environmental Education Research.* ESRC & ERTCEE, Bradford.

Leal Filho W.D.S. & MacDermott, F.D.J. (Eds.) (1995) *Report of the European Seminar on Environmental Education Research.* ESRC, Commission of the European Communities & ERTCEE. ERTCEE, Bradford.

Leal Filho W.D.S., MacDermott, F.D.J. & Murphy, Z. (Eds.) (1995) *Practices in Environmental Education in Europe.* ERTCEE, Bradford.

Leal Filho, W.D.S. (1995a) 'Moving towards Europe: the environmental education challenge', paper presented at the *2nd National Training Workshop on Environmental Education in Malta, Valetta, Malta 26-27 May 1995.*

Leal Filho, W.D.S. (1995b) 'The role of environmental education in the training of scientists on global environmental changes in developing countries' *In NATO ASI Series.* Vol. 1 (29), pp. 71-79. Springer Verlag, Berlin.

Meadows, D. (1990) *Harvesting One Hundredfold.* UNEP, Kenya.

National Milieubeleidsplan (1989) *Political Plan for the Environment from the Dutch Government.* Staatsuitgeverig, The Hague.

Newhouse, N. (1991) 'Implications of attitude and behaviour research for environmental conservation', *Journal of Environmental Education.* Vol. 22, pp. 26-32.

Novo, M. (Ed) (1993) *Basis for a Spanish Strategy on Environmental Education (Bases para una Estrategia Espanola de Educacion Ambiental).* ICONA, Madrid.

Orr, D. (1992) *Ecological Literacy: Education and the Transition to a Postmodern World.* State University of New York Press, Albany.

Porritt, J. (1991) 'A vision for the future' In *Policies for Environmental Education and Training 1992 and Beyond.* English Nature, Peterborough, pp. 65-68.

Scott, D. & Willits, F. (1994) 'Environmental attitudes and behaviour: a Pennsylvania survey', In *Environment and Behaviour.* Vol. 22, pp. 239-260.

Stapp, W., Caduto, M., Mann, L. & Novak, P (1980) 'Analysis of pre-service environmental education of teachers in Europe and an instructional model for furthering this education', *Journal of Environmental Education.* Vol. 12, pp. 3-10.

Tilbury, D. (1992) 'Environmental education within pre-service teacher education: the priority of priorities', In *International Journal of Environmental Education and Information.* Vol. 11, pp. 267-9.

UNEP, IUCN, WWF (1990) *Caring for the Earth: A Strategy for Sustainable Living.* UNEP, IUCN, WWF, Gland.

UNESCO-UNEP (1990) 'Environmentally educated teachers: the priority of priorities?' *Connect* XV (1), pp. 1-3.

WCED (1987) *Our Common Future: Report of the World Commission on Environment and Development.*, WCED, Oxford University Press, Oxford.

INDEX

Printed and bound by CPI Group (UK) Ltd, Croydon, CR0 4YY

23/10/2024

01777667-0001